Co[illegible]

MW01711675

Hidden Dangers

ENVIRONMENTAL CONSEQUENCES OF PREPARING FOR WAR

ANNE H. EHRLICH AND JOHN W. BIRKS, EDITORS

SIERRA CLUB BOOKS ♦ SAN FRANCISCO

Dedication: for LuEsther

The Sierra Club, founded in 1892 by John Muir, has devoted itself to the study and protection of the earth's scenic and ecological resources—mountains, wetlands, woodlands, wild shores and rivers, deserts and plains. The publishing program of the Sierra Club offers books to the public as a nonprofit educational service in the hope that they may enlarge the public's understanding of the Club's basic concerns. The point of view expressed in each book, however, does not necessarily represent that of the Club. The Sierra Club has some sixty chapters coast to coast, in Canada, Hawaii, and Alaska. For information about how you may participate in its programs to preserve wilderness and the quality of life, please address inquiries to Sierra Club, 730 Polk Street, San Francisco, CA 94109.

Copyright © 1990 by Anne H. Ehrlich and John W. Birks

All rights reserved under International and Pan-American Copyright Conventions. No part of this book may be reproduced in any form or by any electronic or mechanical means, including information storage and retrieval systems, without permission in writing from the publisher.

Library of Congress Cataloging in Publication Data
Hidden dnagers : environmental consequences of preparing for war / by
Anne H. Ehrlich & John W. Birks, editors.
p. cm.
ISBN 0-87156-670-2
1. Nuclear weapons plants—Environmental aspects—United States.
2. United States—Armed Forces—Environmental aspects. I. Ehrlich, Anne H.
II. Birks, John W.
TD195.N85H53 1990 90-33841
363.17'99—dc20 CIP

Cover design by Lynne O'Neil
Book design by Seventeenth Street Studios
Typesetting by BookPrep
Printed in the United States of America on acid-free paper
containing a minimum of 50% recovered waste paper, of which
at least 10% is post-consumer waste
10 9 8 7 6 5 4 3 2

Acknowledgments

SEVERAL OF OUR COLLEAGUES AND ASSOCIATES REVIEWED CHAPTERS OF THIS book for us, and we deeply appreciated their help. They are: Stephen Schneider (National Center for Atmospheric Research); Ward B. Watt and Paul R. Ehrlich (Dept. of Biological Sciences, Stanford University); Shirley Feldman (Psychology Dept., Stanford University); Tom Cochran (Natural Resources Defense Council); Robert Ornstein (University of California Medical School, San Francisco). We also wish to thank John Holdren, Donald Lunde, and others who helped us by suggesting potential authors for the chapters.

Our editor at Sierra Club Books, Danny Moses, has given us much help and encouragement along the way. At Stanford University, Steve Masley and Pat Browne patiently photocopied reams of chapter materials in the reviewing and editing phases of the book, and the staff of Falconer Library expeditiously as always produced numerous needed references. At the University of Colorado, several people were helpful in organizing the small conference in October 1988, which was a part of this project, and we wish to thank them. Cheryl Brazeau provided valuable accounting assistance; Lynn Walloch arranged the facilities; Jeanette Trebing provided advertising and travel arrangements for the speakers; and Tim Miller organized the video taping. We also are grateful to the members of Sierra Club's Committee on Military Impacts on the Environment for their encouragement and help with this project.

Finally, we wish to thank the MacArthur Foundation for its grant, which enabled us not only to assemble a far better book than would otherwise have been possible but also to sponsor a small conference with presentations by each of the book's authors. Since the latter represent a wide variety of disciplines and viewpoints, the opportunity for them to exchange information and ideas proved of considerable value to both the audience and themselves.

Introduction: An Unprotected Populace

SAY THE WORD *DEFENSE* AND MOST PEOPLE THINK OF PROTECTION FROM AN external threat: the danger of attack from another nation. Since the end of World War II, two generations ago, the United States and the Soviet Union have warily regarded each other as deadly enemies and built up monstrous arsenals of potent conventional, unconventional, and nuclear weapons. By the 1970s, each nation's nuclear arsenal was capable of not only annihilating the other nation many times over, but of causing such devastation to the entire world as to make the survival of civilization problematic. Yet the weapons continued to pile up as technical advances made them ever more strategically threatening. The world's attention has understandably been fastened on the strategic implications and the rising risks of **using** these growing arsenals.

Concern quite properly has also been expressed about the potential for proliferation of nuclear weapons technology to other powers, especially to certain nations whose governments are often considered less stable and responsible than those of the major powers. Another worrisome trend has been the mounting arsenals of ever more deadly "conventional" arms in developing nations. By diverting needed resources from economic development, this buildup has fueled numerous brutal wars and internal struggles, with far from negligible environmental costs. These relatively minor conflicts of course also enhance the chances of triggering an accidental nuclear war.

Yet, overlooked in the dynamics of the arms races—and their possible culmination in a nuclear holocaust—has been the comparatively less horrendous environmental consequences of making, assembling, transporting, and deploying modern weapons, including mining, refining, and processing the raw materials and disposing of the eventual wastes.

In the United States, the lack of public attention until recently to any untoward side-effects of weapons-building has been abetted by the shroud of secrecy that has always surrounded matters of "national defense." If complaints were heard or questions raised, government agency spokespeople wrapped themselves in the flag and declared that the noble goal of defending the American people surely more than justified a few minor incidents of pollution or risks borne by the well-paid workers in defense industries. The people's right to know did not extend to security affairs. Often the same people who were causing the damage were allowed to define the limits of "security."

Perhaps not surprisingly, "national security" has also been invoked to exempt American weapons-making programs from numerous laws and regulations that apply to industry and all other government agencies: laws to protect the environment and human health, including that of workers. The Department of Energy, responsible for producing America's nuclear weapons, has claimed that it was self-regulating and refused to submit to regulation by other agencies or by the states.

Assertions of threats to national security have clearly hindered claims against the government for damages or injuries inflicted by accidents or improper use of weapons materials, from the sheep killed by the accidental release of nerve gas in Utah in the 1950s to exposures of military personnel and civilians to radiation during the years of above-ground nuclear bomb testing or to Agent Orange in Vietnam. Quite likely, many claims are true and deserve compensation. But the government, faced with a potential multibillion-dollar compensation bill for injuries to tens of thousands of veterans and not a few civilians—which could flush any hope of correcting the budget deficit and might open the door to a flood of new claims—has rejected virtually all of them.

The budget deficit itself, of course, is in part the result of an accelerated military weapons buildup started by President Carter and inflated further during the eight-year Reagan administration. The growth of the deficit after 1981 almost uncannily matched the increases in the defense budget. These increases paid for the accelerated production of weapons and development of new weapons systems, as well as expanded activity by the military in general. Thus, not only has the American public been saddled with a huge debt for the foreseeable future, possibly mortgaging the health and productivity of the nation's economy into the next century, but also it has endured an increase in indirect costs, especially environmental costs.

The hidden environmental costs of the United States military and weapons-building programs are now coming to light, and the public is discovering, to its growing horror, that these costs are not at all inconsequential. Indeed, it may be only in comparison with an actual nuclear war that they could be considered "minor." Compounding the environmental effects that would inevitably accompany even a carefully operated weapons program, appalling abuses and cover-ups have continued to characterize the nuclear weapons-making program, long after the hazards of exposure to radioactive and highly toxic chemicals were clearly understood.

Even if not another warhead is produced, Americans now face a need to clean up the environment and safely dispose of radioactive and toxic chemical wastes on a scale that dwarfs even the civilian Superfund program. The total damage to the environment and to the lives of local citizens has yet to be fully tallied. Nor have all the effects yet been uncovered of past programs of producing chemical and biological weapons—which also require a major clean-up. Numerous U.S. citizens and their offspring are living with the aftereffects of wartime exposure to ecocidal chemicals in Vietnam and peacetime exposure to radiation from production and testing of weapons.

If the costs are counted accurately, we may well wonder: Can the United States afford to build and maintain an overwhelming arsenal of the kind it now has, to say nothing of embarking on a new arms race, as "third-generation" nuclear weapons, the SDI program, or renewed research and development of chemical or biological weapons could commit us to do? Would our children, who will have to pay the price, thank us for the obligation? Does any observable military threat to our security justify such a burden? Indeed, can the world afford the supermilitarization of superpowers and everyone else? Read on . . .

Anne H. Ehrlich and John W. Birks
May 1989

PART ONE

The Nuclear Legacy

1

The U.S. Nuclear Warhead Production Complex

THOMAS B. COCHRAN

DAN W. REICHER

JASON SALZMAN

NUCLEAR WARHEADS ARE DESIGNED AND BUILT BY THE U.S. DEPARTMENT OF Energy (DOE) and dozens of corporations and universities at 17 major sites in 13 states. DOE's nuclear weapons production facilities occupy twice as much land as the state of Delaware and employ more than 90,000 people (see map). Since the 1940s, the weapons complex has produced some 60,000 warheads, over one-third of which are currently in the arsenal.

DOE's 1988 nuclear weapons production budget—including funds for nuclear weapons design, testing, assembly, and disassembly, as well as nuclear material and component production—was about $8 billion. In 1988 dollars, this exceeds spending during the Manhattan Project and approaches the record-high budgets for nuclear weapons production of the late 1950s and early 1960s. During more than 40 years of weapons production, the United States has spent about $104.5 billion ($250 billion in 1988 dollars) on the manufacture of nuclear warheads and another $800 billion ($1.95 trillion in 1988 dollars) on delivery systems and other support costs. An additional $100 billion or more may be necessary for environmental cleanup at the weapons complex.

The Energy Department's expenditures for warhead production

activities increased by an average of 18 percent annually from fiscal year 1978 through fiscal year 1982. Since fiscal year 1983, DOE's defense budget has increased at a slower rate, on an average of less than 10 percent annually. About 60 percent of DOE's total budget is allocated to weapons production, up from about 25 percent in fiscal year 1978. DOE's non-defense work includes research and development in areas such as nuclear power, fossil fuels, solar energy, and energy conservation.

The first U.S. nuclear weapons were developed by the U.S. Army's Manhattan Engineering District, or "Manhattan Project," during World War II. Following the war, control of all aspects of atomic energy was turned over to the Atomic Energy Commission, established by the Atomic Energy Act of 1946. Reflecting concerns about the use of atomic energy, the act limits private ownership of fissionable material and classifies as "restricted data" all information relating to the development of nuclear weapons. The act also provides for civilian rather than military control over the development of nuclear warheads and naval reactors.

In 1974, Congress split the Atomic Energy Commission into the Nuclear Regulatory Commission (NRC), established to regulate commercial applications of nuclear energy, and the Energy Research and Development Administration (ERDA), in charge of nuclear weapons

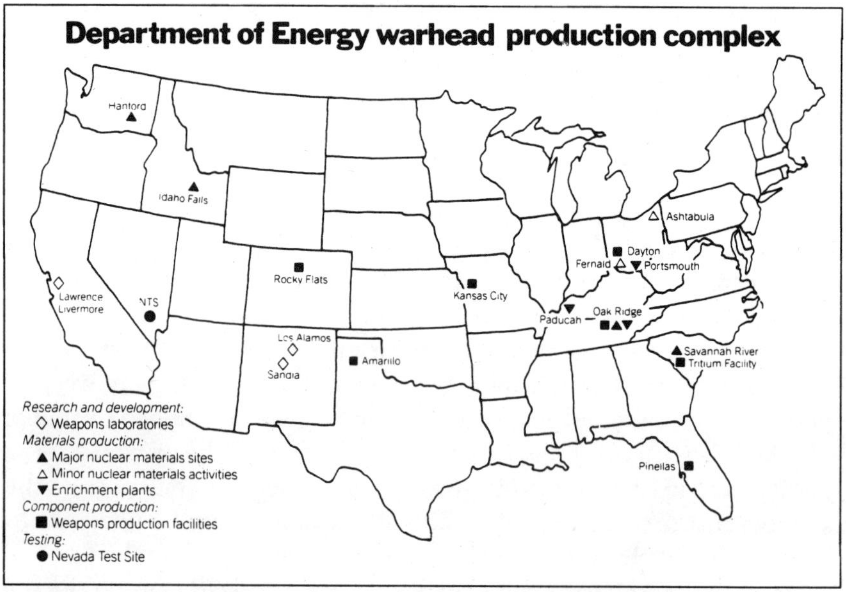

Figure 1-1: Department of Energy warhead production complex

activities and research relating to energy technology. In 1977, under President Carter ERDA became the Department of Energy.

Although the weapons facilities are owned by the federal government, the plants are operated by more than 70 contractors that bid for DOE work. The contractors at the government-owned contractor-operated (GOCO) facilities include some of the largest U.S. corporations, such as General Electric, Monsanto, and Westinghouse, as well as major universities including the University of California and the University of Chicago. (see Table 1).

Facility	Current Operating Contractor	Date of Initial Operation
RESEARCH LABORATORIES		
Los Alamos Laboratory (LANL) Los Alamos, New Mexico	University of California	1947
Lawrence Livermore National Laboratory (LLNL) Livermore, California	University of California	1952
Sandia National Laboratories (SNL) Albuquerque, New Mexico Livermore, California (SNLL)	Sandia Corporation, subsidiary of AT&T Corporation	1945
MATERIALS PRODUCTION FACILITIES		
Feed Materials Production Center Fernald, Ohio	Westinghouse Materials Co. of Ohio	1953
Ashtabula Plant Ashtabula, Ohio	Westinghouse Materials Co. of Ohio	1952
Y-12 Plant Oak Ridge, Tennessee	Martin-Marietta Energy Systems, Inc.	1944
Hanford Reservation Richland, Washington	Westinghouse Hanford Co.	1944
Savannah River Site Aiken, South Carolina	Westinghouse Savannah River Company	1950

Table 1: Major DOE Facilities and Operating Contractors

Facility	Current Operating Contractor	Date of Initial Operation
Idaho National Engineering Laboratory (INEL) Idaho Falls, Idaho	Westinghouse Idaho Nuclear Corporation and EG&G Idaho, Inc.	1949
Paducah Gaseous Diffusion Plant Paducah, Kentucky	Martin-Marietta Energy Systems, Inc.	1954
Portsmouth Gaseous Diffusion Plant Piketon, Ohio	Martin-Marietta Energy Systems, Inc.	1956
WARHEAD PRODUCTION FACILITIES		
Rocky Flats Plant Golden, Colorado	EG&G, Inc. (as of January 1990)	1951
Y-12 Plant Oak Ridge, Tennessee	Martin-Marietta Energy Systems, Inc.	1944
Savannah River Plant Aiken, South Carolina	Westinghouse	1950
Mound Facility Miamisburg, Ohio	EGAG Applied Technologies	1948
Pinellas Plant St. Petersburg, Florida	General Electric Co.	1957
Kansas City Plant Kansas City, Missouri	Allied-Signal Aerospace Co.	1949
Pantex Plant Amarillo, Texas	Mason & Hanger—Silas Mason Co., Inc.	1951
TEST SITES		
Nevada Test Site Nye County, Nevada	Reynolds Electrical & Engineering Co.; Edgerton, Germeshausen, & Grier, Inc.; Holmes & Narver, Inc.; Fenix & Sisson	1951
Tonopah Test Range Nye County, Nevada	Reynolds Electrical and Engineering Company (site service)	1957

Table 1: Major DOE Facilities and Operating Contractors continued

The Energy Department's nuclear weapons production activities and its budget requests to Congress are guided by the Administration's annual "Nuclear Weapons Stockpile Memorandum." The classified memorandum is drafted by officials from the departments of Energy and Defense, reviewed by the National Security Council, and approved by the President. It outlines the present and future composition and size of the nuclear warhead stockpile. DOE produces nuclear materials and warheads in accordance with the requirements outlined in the memorandum.

Technically, the Department of Energy does not produce nuclear weapons—only the warheads that are attached to those weapons. A nuclear warhead contains an explosive nuclear or thermonuclear system. The delivery systems for nuclear warheads are produced by the Department of Defense.

Four congressional committees and four associated subcommittees are primarily responsible for DOE's nuclear warhead production activities: the Senate Appropriations Committee, Subcommittee on Energy and Water Development; the Senate Armed Services Committee, Subcommittee on Strategic and Theater Nuclear Forces; the House Appropriations Committee, Subcommittee on Energy and Water Development; and the House Armed Services Committee, Subcommittee on Procurement and Military Systems.

NUCLEAR WEAPONS PRODUCTION ACTIVITIES

In the following four sections, we describe the major activities associated with nuclear weapons production: research and design, nuclear materials production, warhead component production, and warhead testing. We conclude the chapter with a brief history of the growth of the complex and a section about DOE's deteriorating weapons production facilities.

Research and Design

Most nuclear warhead research and design takes place at three DOE laboratories: Los Alamos National Laboratory in New Mexico, Lawrence Livermore National Laboratory in California, and Sandia National Laboratories in New Mexico and California. Two smaller DOE laboratories—the Savannah River Laboratory in South Carolina and the New Brunswick Laboratory in Illinois—and several DOE multipurpose laboratories around the country also engage in some research related to nuclear weapons. Army, Navy, and Air Force laboratories supplement these activities, conducting research on delivery systems, nuclear effects, and weapons safety.

Although nuclear warheads are designed in numerous ways, most or all of the explosive energy for every warhead is derived from fission, fusion, or a combination of the two nuclear processes.

Fission Weapons. When a nucleus of a uranium or plutonium atom splits (fission), energy and neutrons are released. The energy from a fission reaction is about ten million times as much, atom for atom, as is obtained from ordinary chemical combustion. To create a fission explosion in a typical weapon, a very large number of atomic nuclei must be fissioned within a short period of time, typically a few microseconds. This is accomplished by initiating an "uncontrolled chain reaction," in which neutrons released from one "generation" of fissioning nuclei rapidly initiate the fission process in the next generation, and so on.

In a typical fission weapon, a chemical high explosive initiates the chain reaction by compressing or pushing together the "fissile materials"—plutonium (primarily plutonium 239) or highly enriched uranium (highly concentrated uranium 235)—into a "supercritical mass." Once a supercritical mass is formed, energy and neutrons are released, which split uranium and plutonium atoms. A chain reaction occurs, which fuels the subsequent explosion.

Fusion Weapons. In a thermonuclear or "hydrogen" explosion, hydrogen isotopes deuterium and tritium—called "thermonuclear fuel"—are fused to form helium, releasing one neutron and considerable energy. It is not necessary, however, to use solely elemental deuterium and tritium directly in a thermonuclear weapon. Instead, tritium can be produced inside the weapon itself. In this process, lithium 6, in the chemical form of lithium-deuteride, is bombarded by neutrons during the fusion reaction to produce the needed tritium.

The fusing of thermonuclear fuel will occur only at a temperature in the range of tens of millions of degrees. In order to achieve the required temperature and pressure in fusion bombs, a small fission device is set off first, which initiates the fusion reaction. In fusion bombs, the fission reaction is referred to as the "primary" stage, or "trigger," and the fusion reaction is called the "secondary" stage. The radiation from the fusion process in the secondary stage can be contained and used to transfer energy to compress and ignite a "tertiary" stage, and that stage could similarly ignite a fourth stage, and so on. There is no theoretical limit to the number of stages that might be used and, consequently, no theoretical limit to the size and explosive power, or "yield," of a thermonuclear weapon. The fission of uranium 238 atoms is often used in the final stage to increase significantly the yield of the weapon.

Deuterium and tritium are often added to fission warheads and the fission primaries of thermonuclear warheads to provide an additional source of neutrons to speed up the chain reaction. This increases or "boosts" the yield of the warhead, thereby enabling the construction of lighter and smaller nuclear warheads. Deuterium and tritium are also essential ingredients in enhanced radiation warheads, commonly called "neutron" bombs.

Nuclear Materials Production

Of the six principal nuclear materials used in nuclear weapons—uranium 235, uranium 238, plutonium 239, tritium, deuterium, and lithium 6—only uranium 238 is abundant in nature. Plutonium and tritium are produced in nuclear reactors, called "production" reactors. The concentration of U-235 in natural uranium is increased in a set of special "enrichment" facilities. Deuterium, in the form of "heavy water," is produced in heavy-water production plants. In other facilities, lithium 6 is separated from naturally occurring lithium. All these materials except tritium can be stockpiled for extended periods.

In recent years, only plutonium and tritium have been produced. The requirements for other nuclear materials are met from existing inventories. No lithium for weapons has been produced since the early 1960s. President Johnson halted the production of highly enriched uranium for weapons in 1964, and production of deuterium ceased in 1982. Gold, beryllium, aluminum, chromium, nickel, tin, titanium, and tungsten—nonnuclear materials used in nuclear warheads—are acquired by the government from the National Defense Stockpile of Strategic and Critical Materials.

The steps required for the production of nuclear materials, including transportation and waste disposal, are collectively termed the "fuel cycle." In theory, this cycle starts with the mining of naturally occurring uranium and ends with the final assembly of nuclear weapons and the disposal of wastes. In practice, the processes required for nuclear warhead production do not necessarily proceed in such a predictable, step-by-step manner. Most nuclear materials for new warheads are obtained from retired warheads rather than new production. These materials are recovered, purified, and processed before re-use. Also, at a number of DOE facilities, nuclear materials are recovered from spent nuclear fuel and scrap generated from nuclear weapons production, rather than being produced from raw materials. However, a description of the complete nuclear fuel cycle is helpful for understanding the many facets of nuclear weapons production—and the resulting environmental impacts.

Uranium Mining and Milling. The fuel cycle begins with the mining of uranium ore. In 1948, the Atomic Energy Commission initiated a program of incentives that included price guarantees and bonuses for discovery and delivery of ore in order to spur private uranium mining ventures. The response was so great that the government was able to terminate domestic natural ore purchases in 1962.

Once mined, raw uranium ore must be chemically processed or "milled," which typically increases the uranium oxide content from 0.1 percent in the ore to 85 to 95 percent in the milling product, called yellowcake. Prior to 1971, when the United States stopped purchasing yellowcake, the government obtained about 55 percent of the material for weapons production from more than 25 mills in New Mexico, Wyoming, Utah, and other states, and the remainder from foreign sources, primarily Canada and Zaire. The milling wastes, called "tailings," were deposited in piles near the milling facility. These tailings, consisting of radium and its daughter products, are radioactive.

Uranium Enrichment and Fuel Fabrication. The concentration of uranium 235 in yellowcake is 0.7 percent. Before uranium 235 can be used for weapons production plants, its concentration must be increased in a process called enrichment. An enrichment facility at Paducah, Kentucky, increases the uranium 235 concentration to about 2 percent. From there it is sent to an enrichment plant at Portsmouth, Ohio, where uranium 235 is further concentrated.

Enriched uranium has a variety of applications. Enriched to about 1 percent, it was previously used to produce plutonium at production reactors at the Hanford Reservation in Washington State; enriched to about 3 to 4 percent, it fuels commercial light water power reactors; enriched to about 20 percent or greater, it fuels research and test reactors; enriched to about 60 percent it is used to produce plutonium and tritium at production reactors at the Savannah River Plant in South Carolina; enriched to 93.5 percent, it is used to make components of nuclear warheads;[1] and enriched to 97.3 percent, it fuels U.S. submarine reactors.

Before highly enriched uranium can be used in a production reactor, it is processed and converted to metal at the Y-12 plant in Oak Ridge, Tennessee; the metal is fabricated into "fuel rods" at the Savannah River Plant.

Uranium 238, or "depleted uranium," left over from the enrichment process is sent to DOE plants at Fernald and Ashtabula, Ohio, where it is converted into uranium metal and extruded into "target rods" for plutonium production. Target rods for tritium production are manufactured from lithium 6 at the Savannah River Plant.

The target rods and fuel rods are placed in production reactors at the department's Savannah River Plant. In a production reactor, a fission chain reaction, similar to that which occurs in a nuclear warhead, is initiated in the fuel rods. The chain reaction in a production reactor, however, is slowed by "control rods." These rods are inserted amidst the fuel rods to absorb neutrons and thereby control the rate of fission reactions. While some neutrons generated during the controlled fission process perpetuate the chain reaction, others are absorbed by the nuclei of uranium 238 or lithium 6 contained in the target rods. When an atom of uranium 238 absorbs a neutron, it is transformed into plutonium. When an atom of lithium 6 absorbs a neutron, it is transformed into tritium.

Chemical Separation. Unwanted and highly radioactive "fission products"—materials such as iodine 131, strontium 90, and cesium 137—are also produced during the fission chain reaction. Before being used in warheads, plutonium must be separated from the unwanted fission products in "chemical separation" facilities. In such plants, which DOE operates at the Savannah River Plant and the Hanford Reservation, powerful acids dissolve irradiated fuel and target rods and the plutonium is extracted. Irradiated lithium target rods are heated to drive out tritium, which is collected as a gas. This operation is performed at the Tritium Facility at the Savannah River Plant.

---

[1] As noted above, DOE has not produced highly enriched uranium for nuclear warhead production since 1964. Instead, DOE relies on an existing stockpile of the material and also recovers it from the spent fuel from nuclear-powered submarine reactors and research reactors at "chemical separation" facilities at the Idaho National Engineering Laboratory and the Savannah River Plant. DOE also uses the chemical separation facility at the Savannah River Plant to extract enriched uranium from the spent fuel from production reactors.

Defense Waste. DOE manages wastes resulting from weapons production in various ways. Highly radioactive liquid wastes, which contain long-lived and/or highly radioactive materials, are placed in underground steel tanks at the Hanford Reservation and the Savannah River Plant. At Hanford, more than 55 million gallons (more than 470 million curies) of high-level waste are stored in 169 steel tanks, and at the Savannah River Plant, more than 29 million gallons (more than 775 million curies) of high-level waste are stored in 51 steel tanks. Corrosion and cracking of some of the tanks have led to the release of radioactive materials into the soil. At Hanford, for example, hundreds of thousands of gallons of highly radioactive wastes have leaked into the ground from the tanks.

At the weapons plants, solid low-level wastes have been buried in unlined pits, and liquid low-level wastes, often containing nonradioactive toxic chemicals, have been disposed of in unlined "seepage basins." The term "low-level radioactive waste" means radioactive waste not classified as high-level waste, transuranic waste, spent nuclear fuel, or by-product material, as those terms are defined by the Atomic Energy Act. At some weapons plants, these wastes have seeped into groundwater.

In December 1988, DOE estimated that cleanup of its facilities would cost between $42 billion and $72 billion, not including day-to-day "base program" activities. The General Accounting Office subsequently estimated that the cost would be $80 billion to $110 billion—about $1.3 million to $1.8 million for each warhead the United States has produced.

Warhead Component Production

After nuclear warhead materials are produced, they are fabricated into specific components. Warhead components are manufactured at three facilities. At the Rocky Flats Plant near Denver, Colorado, plutonium is made into fission triggers for the fusion reaction in hydrogen bombs. Facilities at Oak Ridge, Tennessee, manufacture uranium 235 components for fission triggers, and the lithium, deuterium, and uranium 238 components of thermonuclear weapons. Tritium is loaded into containers called "reservoirs" at the Savannah River tritium facility for subsequent insertion into weapons.

Nonnuclear components of nuclear warheads—timers, detonators, rubber, plastic, and foam—are produced at three other facilities. The

Mound Laboratory in Miamisburg, Ohio, makes detonators and various parts of the firing circuits. The Pinellas Plant in St. Petersburg, Florida, manufactures neutron generators, devices that initiate the fission chain reaction in weapons. The Kansas City Plant in Missouri makes electronic and other nonnuclear parts.

These nuclear and nonnuclear components are shipped to the Pantex Plant near Amarillo, Texas, where final assembly of the warheads takes place. The warheads are then turned over to the Department of Defense for deployment around the globe.

Warhead Testing

The first U.S. nuclear test occurred on July 16, 1945, on a 100-foot tower at the White Sands Bombing Range, 55 miles northwest of Alamogordo, New Mexico. As of the end of 1988, the United States had conducted about 920 known tests, an average of about one test every 17 days. Of these, about 790 took place at the Nevada Test Site near Las Vegas, Nevada; 108 in the Pacific; 17 in various other parts of the United States, including Alaska; and 3 in the South Atlantic. Over 100 tests were not announced by the U.S. government and were subsequently discovered upon examination of seismic data. The vast majority of U.S. nuclear tests were conducted for military purposes. They have occurred atop towers, on barges, suspended from balloons, underwater, and underground. Today, nuclear testing is confined to the Nevada Test Site. As stipulated by the 1963 Limited Test Ban treaty, all explosions now take place underground.

Most U.S. nuclear tests are categorized in one of four ways: weapons related, weapons effects, safety experiment, and "plowshare." Weapons related tests, which constitute over two-thirds of the U.S. total, are conducted to develop new warhead designs, to certify the yield and military characteristics of new warheads, to obtain information about the fundamental phenomena of a nuclear explosion, or, occasionally, to insure that an existing warhead type in the arsenal still performs as expected. Weapons effects tests are undertaken to gather data on the air blast, ground and water shock, heat, electromagnetic pulse, neutrons, and gamma and X rays resulting from a nuclear explosion. This information is used to assess the survivability of U.S. military systems in a nuclear war environment and to predict the impact of the explosion on enemy forces and equipment. Safety experiment tests are designed to confirm that a nuclear explosion will not occur in case of an accidental detonation of the chemical high explosive associated with the device. About 25 plowshare tests were conducted between 1961 and 1973 to develop peaceful uses of the atom.

During most nuclear tests, the device to be tested is placed in a cylindrical capsule or canister, which is inserted into a vertical shaft 600 to 5,000 feet deep. The shaft is filled with gravel and epoxy plugs to prevent the escape of radioactive materials. Upon detonation, the test device creates a large underground cavity, the bottom of which quickly fills with molten rock and debris. The cavity often collapses, creating a large crater at the surface. The Nevada Test Site is pockmarked with several hundred craters of various sizes, from 200 to 2,000 feet in diameter and up to 200 feet deep. Astronauts have used the test site for training missions prior to their journeys to the moon.

At the Tonopah Test Range near the Nevada Test Site, warheads loaded with test equipment and nonexplosive substances are used to test the ballistics of bombs, artillery shells, and rockets. Defense Department test ranges in Florida, California, and New Mexico supplement these facilities.

WARHEAD PRODUCTION RATES

From 1955 to 1967, some 30,000 new nuclear warheads were produced. The highest rate of warhead production occurred between 1959 and 1961, when 5,000 to 6,000 warheads were manufactured each year. The nuclear arsenal reached its numerical high of about 32,500 warheads in 1967, when 17 different types of warheads were in production simultaneously.

After the enormous warhead buildup of the 1950s and 1960s, the size of the stockpile stabilized, and then began to decrease. During most of 1977 and part of 1978, for example, only a few hundred warheads of just one design were produced. Beginning in 1978, however, activity at the production complex increased. The waning of detente, the Soviet invasion of Afghanistan, the U.S. failure to ratify SALT II, and the election of Ronald Reagan in 1980 influenced decisions to increase production. New weapons on the drawing board or in production included the Cruise, MX, Trident I, and Pershing II missiles, as well as neutron weapons and other new bombs.

The development of these new weapons raised concerns about whether enough nuclear materials would be available for their manufacture. During 1977 and 1978 members of the Senate and House Armed Services committees visited Energy Department facilities and found that the department lacked a comprehensive program to meet the pending warhead production schedule. Executive branch committees were formed to examine the situation. In June 1979, DOE submitted a report to Congress that identified deficiencies in the production complex and provided a five-year plan to correct them. A policy review committee directed by the National Security Council concluded in 1980 that the nuclear material production capacity should be expanded. A joint Department of Defense/Department of Energy group was directed to develop plans to upgrade and expand the production complex over the next 20 years.

Upon entering office, President Reagan initiated the production of smaller, more accurate warheads requiring more plutonium and tritium, driving projected nuclear materials requirements higher. In approving his second Nuclear Weapons Stockpile Memorandum in November 1982, Reagan called for the creation of "sufficient reserves" of nuclear materials. These reserves were said to be needed "as insurance against unforeseen [nuclear materials] production interruptions and to allow for [a warhead production] surge capacity."

To meet higher production goals, Reagan increased the warhead production budget beyond the increases that were made during the last two years of the Carter administration. From 1981 to 1986, plutonium production nearly doubled and the capacity of the weapons complex was expanded. By 1986, after a flurry of retirements and new production, about one-third of the warheads in the arsenal were under six years old. But at the same time—contrary to widespread perception—the total number of warheads in the nuclear arsenal decreased from 24,000 in 1981 to 23,400 in 1986. This downward trend has continued, with about 21,000 warheads in the stockpile today.

AGING WEAPONS FACILITIES

Increased activity at the warhead production complex in the late 1970s and 1980s came at a time when many DOE weapons facilities were beginning to show signs of age, raising serious concerns about their safety.

All of DOE's major facilities are over 25 years old. Most were built in the early 1950s and some still in operation today date back to the 1940s. The Y-12 and Rocky Flats plants, for example, began operating in 1944 and 1951, respectively.

Of the 14 reactors built to produce materials for nuclear weapons, only 3 at the Savannah River Plant have operated recently. The eight original graphite production reactors at the Hanford Reservation were shut down between 1964 and 1971; the "R" reactor at the Savannah River Plant was placed on "standby" in 1964 and is now permanently shut down; the "C" reactor at the Savannah River Plant was shut down indefinitely for safety reasons in July 1985.

In January 1987, DOE temporarily shut down the "N" reactor at the Hanford Reservation, after a panel of outside experts raised concerns about its safety in the aftermath of the Chernobyl disaster. The "N" reactor shares some similarities with the stricken Soviet plant. In February 1988, citing a decreased need for plutonium, the department placed the reactor on "standby" status for use only in an emergency.

The three reactors that have operated recently—the "P," "K," and "L" at the Savannah River Plant—were all built in the early 1950s. (The L reactor was placed on standby in 1968 and restarted in 1985.) All three reactors were shut down during 1988, pending the resolution of safety concerns.

With major weapons plants shut down or in serious disrepair, and with massive environmental problems at many sites, DOE has proposed to construct new facilities, renovate existing ones, and clean up the weapons complex over the next several decades at a cost that could exceed $150 billion.

REFERENCES

Thomas B. Cochran, William M. Arkin, and Milton M. Hoenig, *Nuclear Weapons Databook Volume I: U.S. Nuclear Forces and Capabilities*, Ballinger, Cambridge, Mass., 1984.

Thomas B. Cochran, William M. Arkin, Robert S. Norris, and Milton M. Hoenig, *Nuclear Weapons Databook Volume II: U.S. Nuclear Warhead Production*, and *Nuclear Weapons Databook Volume III: U.S. Nuclear Warhead Facility Profiles*, Ballinger, Cambridge, Mass., 1987.

Department of Energy (DOE), "Environment, Safety and Health Report for the Department of Energy Defense Complex," July 1, 1988, Washington, D.C.

National Research Council, "Safety Issues at the Defense Production Reactors: A Report to the Department of Energy," National Academy Press, Washington, D.C., 1987.

Robert S. Norris and William Arkin, "Nuclear Notebook," *Bulletin of the Atomic Scientists,* March 1988, p. 56. Also see Robert S. Norris, Thomas B. Cochran, and William Arkin, "Known U.S. Nuclear Tests, July 1945 to 31 December 1987," Nuclear Weapons Databook Working Paper 86-2 (rev. 2), available from the Natural Resources Defense Council, Washington, D.C.

2

The Generation Time-Bomb: Radioactive and Chemical Defense Wastes

MARVIN RESNIKOFF

THE DESIGN AND PRODUCTION OF NUCLEAR WEAPONS BEGAN UNDER GREAT secrecy at the end of World War II. Under wartime emergency conditions, nuclear weapons had to be designed and tested rapidly. After the weapons were proven a "success" in Japan, the original production facilities at Oak Ridge and Los Alamos were expanded into a gigantic nationwide production complex.

Under wartime emergency conditions, concerns about the effects of radiation on workers and the safe storage of the by-products of nuclear weapons production, nuclear waste, were a distant second to meeting production schedules. Liquid high-level waste was poured into single shell tanks at Hanford. Liquid "low-level" and plutonium-contaminated wastes were poured into the ground in "pits" and "cribs" at Hanford, the Savannah River Plant, the Y-12 Plant, and the Feed Materials Production Center near Cincinnati, and directly into the Snake River Aquifer at the Idaho National Engineering Laboratory. Solid low-level waste in cardboard and wood boxes and steel drums was placed in landfills, identical to municipal garbage-type landfills. The pits and cribs were designed to leak underground; the pits were designed only to retard the leakage, allowing the radioactivity to decline somewhat before reaching humans. The Earth is, after all, an imperfect container.

In regions of high precipitation, if liquid waste did not seep below ground, the pits would fill like bathtubs and overflow into surface streams. The radioactive concentrations that eventually reached the public were expected to be low, so low that health effects would be negligible. Because of the secrecy, the wartime urgency, and the fact that the brightest minds in the country were employed by the federal government, it was impossible to question these shoddy practices. In the early days, the federal government controlled all the information.

Waste management was never a criterion when sites were being selected for weapons manufacturing plants. Many Department of Energy sites were located above major aquifers that supplied drinking water to millions of people or fed into major rivers: Savannah River Plant (Tuscaloosa aquifer), Feed Materials Production Center and Mound Laboratory (Great Miami aquifer), Hanford (Columbia River), Pantex (Ogalalla aquifer), Idaho National Environmental Laboratory (Snake River aquifer), and Oak Ridge National Laboratory (Clinch River). In fact, the waste management problem was not recognized as such until the bomb-producing factories were in full production. As Bob Loux, head of the Nevada Nuclear Waste Project, put it, "It's like sending John Glenn into space without any way of bringing him back."

Consider the casual beginnings of waste management at Oak Ridge National Laboratory. This can be traced back to a memorandum, dated January 5, 1944, from the medical department of the Central Safety Committee to Mr. B. Smith, secretary of the Central Safety Committee:

> It would seem that provision has not been made for the disposal of actively contaminated broken glassware, of materials not sufficiently clean to be used in other work. I am suggesting that a metal trash can with cover, with red lettering on the can, be provided for the disposal of such materials as can be placed in the trash can. Mr. Schwertfeger has suggested that a suitable location for the burying of this material could be provided over on the burning ground. A suitable pit with enclosed fence could be made.

With these humble beginnings, burial ground #1 was opened in 1944. From one trash can with red lettering, Oak Ridge National Laboratory now has six burial grounds covering over 200 acres, containing 7 million cubic feet of low-level waste. No records were deemed necessary on the contents of burial grounds #1 and #2, and the early records for burial grounds #3, #4, and #5 were destroyed in a fire in 1961.

Today, a total of 100 million cubic feet of low-level wastes sits in landfills at all the nuclear weapons design and production facilities across the country,[1] enough to fill a container the size of a football field to a height of one-third of a mile.

One hundred million gallons of liquid high-level wastes, with a total activity of 1.4 billion curies, sits in underground tanks at the Savannah River Plant and Hanford and in solid form at the Idaho National Engineering Laboratory. To give some perspective to this quantity of radioactive material, a person can be seriously harmed by one-millionth of a curie of strontium-90, equivalent to an aspirin-sized amount of high-level waste. This waste material will have to be monitored and maintained essentially forever.

Leakage of radioactive materials into the environment inevitably has occurred and will continue to occur. Over 500,000 gallons of high-level wastes have now leaked from the Hanford tanks.[2] Plutonium-contaminated waste, poured into earthen cribs at Hanford, will remain radioactive for tens of thousands of years. At the Savannah River Plant, monitoring wells have shown an increasing spread of strontium-90 from burial grounds and pits into groundwater.[3]

At Oak Ridge National Laboratory, burial grounds are in progressive stages of deterioration. Boxes and steel barrels degrade, waste and trench covers settle. More water enters the trench covers of landfills and leaks out as radioactive liquid. Despite attempts at diverting surface water, strontium-90 continues to leak from burial ground #4, and an underground aquifer surfaces in the middle of it. It is hard to imagine placing a burial ground directly in an aquifer, but such was the cavalier attitude of the Atomic Energy Commission in the 1950s. The National Academy of Sciences has expressed concern[4] that burial grounds #5 and #6 will in time follow the same course as burial ground #4.

Although present burial practices have improved—in that waste is solidified before disposal and carefully rather than randomly stacked, and burial ground covers are carefully mounded to promote surface water runoff—all landfills in high precipitation regions sooner or later leak. At Mound Laboratory and Lawrence Livermore National Laboratory, tritium is leaking off-site. At the Feed Materials Production Center and at the Y-12 Plant, uranium wastes were placed within unlined pits or behind earthen dams. Uranium inevitably leaked out, and this leakage will continue unless the uranium is exhumed.

In semidesert regions, flash floods occur often, creating arroyos and canyons. Plutonium and other radioactive materials have washed off the Los Alamos National Laboratory site into surrounding areas. With little vegetation, wind can blow the covers off landfills, leaving waste exposed. Hanford, Sandia National Laboratory, and the Nevada Test Site are located in dry regions.

At the Nevada Test Site, bombs are tested underground at varying depths within or close to a deep underground aquifer. Though information regarding aquifer movement and systematic well testing is unavailable, it is certain that the aquifer near the underground explosions has become contaminated. It is only a matter of time before contamination reaches the site boundary and the general public. The Nevada Test Site has now become a dump site for waste from many Department of Energy facilities, including thousands of barrels of thorium waste from Fernald. The Oak Ridge National Laboratory, Mound Laboratory, and Lawrence Livermore National Laboratory are now increasingly disposing of materials at the Nevada Test Site.

Officials of the Atomic Energy Commission, and now the Department of Energy, were always quick to defend shoddy practices with several levels of excuses, even though the production facilities were not sited or built with waste management in mind. In regions of high precipitation, such as the Southeast, dilution was their saving grace. According to the Department of Energy, even though a large amount of radioactive waste leaked out, the excess water diluted the radioactivity to safe concentrations. The dose to any individual was therefore presumed low, "almost background." In many cases, it was difficult to know what background was, since no studies of background were done before the plants began operation.

But dilution is not the solution, because although the radiation dose to each person might be less, the number of persons affected would be much greater. The total number of cancers caused is proportional to the total radioactivity ingested by all people. Department of Energy studies have never fully accounted for the effects of radioactive leakage into the environment.

In dry regions, the Department of Energy used another line of reasoning. Here, according to the department, the Earth acted as an ion exchange column, absorbing radionuclides. Radioactive materials were supposed to attach themselves to soil and thereby not enter publicly accessible waters. Since nuclear weapons sites were large, the off-site releases were expected to be small. Little precipitation meant that radiation would stay fixed.

But contrary to expectations, an underground plume of liquid tritium from Hanford has reached and is emptying into the Columbia River. In addition, iodine-129 and plutonium have spread throughout a supposedly confined underground aquifer, more than 13 miles from the likely source and 1,350 feet below ground, again defying Department of Energy predictions. Moreover, in dry areas, where earth-holding vegetation is scarce, wind also has dispersed radioactivity.

Because the sites for plutonium production, the Savannah River Plant and Hanford, were both so large, the Department of Energy argued that radioactive decay would occur before any radioactive material leaked off-site. In the wake of this logic, much radioactive material leaked off-site and major wastelands were created. This has clearly happened at the Savannah River Plant, Hanford, and Oak Ridge.

When off-site radioactive releases rose to embarrassingly high levels, as in the swamp adjacent to the Savannah River Plant, the terms of the debate were shifted once again. The Atomic Energy Commission, now the Department of Energy, fell back on health studies that showed that no health effects would occur at such low doses. Honest scientists who produced health studies at variance with the party line were discharged by the federal agency. The supportive health studies then formed the basis of new regulations. Since low radiation doses were not deemed harmful, the allowable levels of radioactive concentration were set high enough so that radioactive releases from Department of Energy facilities always fell well below them.

If managing waste became too expensive or troublesome, the Department of Energy, with the stroke of a pen, could change the definitions of waste to less restrictive categories. The change in definition of TRU or plutonium-contaminated waste, for instance, from 10 to 100 picocuries of plutonium per liter, led to a reduction of the department's plutonium-contaminated waste from 400 million to 1 million cubic feet.

New regulations also can be used to alter the definition of waste. The Nuclear Regulatory Commission, which sets regulations on high-level waste, has instituted proceedings to alter the definition of high-level waste. Presently, high-level waste is irradiated fuel or the liquids from dissolving this fuel in acid; the NRC would instead define high-level waste by concentration, in effect setting a sliding scale that could later be changed. The proposed change would permit the Department of Energy to retain high-level waste in the 149 single shell tanks at Hanford, saving the federal government over $10 billion in deep underground repository disposal fees. Millions of curies of long-lived toxic materials would be left on the surface of the Earth for the next generation to care for. No health studies accompanied this proposed change.

Honest health statistics researchers in government agencies continue to be fired or leave in disgrace. Health standards are still being relaxed: the allowable amounts of strontium-90 and uranium that the public (to be distinguished from workers) may safely ingest were increased by a factor of ten in 1986.

Of course, the Department of Energy has been self-regulating. It could, in a time of weakness, ask itself if it was doing a good job. The answer would always be yes.

The final line of defense against questions from the American public about shoddy practices, loose regulations, and health studies was always the old standby, national defense. Anyone who questioned the Department of Energy too convincingly was the enemy. Radioactive Waste Campaign researchers (a public-interest environmental organization focussing on radioactive waste issues) who have vigorously questioned shoddy waste management practices at the department's Oak Ridge Reservation were labelled "bomb-throwing researchers."[5]

American corporations would be green with envy at the privileged position of the Department of Energy. Imagine being a major corporation with the right to control the information released to the public, to set the regulations, to police yourself, and, if all else fails, to silence your critics! The effect on the bottom line would be remarkable.

The most ingenious method of waste management is to recycle waste to the general public. What is waste to one person may be gold to another, if sold right. Rather than safely disposing of americium-241 from obsolete nuclear weapons, the department has sold that dangerous alpha emitter for use in smoke detectors. Used smoke detectors are disposed of in municipal landfills. Cesium-137 is removed from high-level waste and used in food irradiation machines. If the cesium-137 leaks from waste capsules, as occurred in an irradiation facility in Georgia in 1988,[6] the low-level waste created is then commercial radioactive waste, which is sent to a commercial landfill. Military high-level waste is thereby converted to commercial low-level waste. Or, as occurred in Brazil, cesium-137 from an irradiation machine was found in a dump and was handled by the curious as a deadly magical powder. The cesium originally came from Hanford. As yet another example, contaminated piping from the Oak Ridge Gaseous Diffusion Plant was sold to a local junkyard and then recycled into other metal products.

Let us examine the radioactive waste situation and the threats posed by each weapons facility across the country:

FERNALD

The Feed Materials Production Center is located on 1,050 acres in Fernald, Ohio, 20 miles northwest of Cincinnati. More than a million people live within 20 miles of the plant. The center processes various uranium compounds into uranium metal, which is used in nuclear weapon components and as fuel and target elements for plutonium production reactors. Over its 34-year history, the Fernald plant has released at least 520,000 pounds of uranium dust into the surrounding air, soil, and water, including thousands of pounds released in the course of numerous accidents that have occurred at the facility. A recent study by the Radioactive Waste Campaign estimates up to 187 additional cancers to local residents may have been caused by inhaling uranium-contaminated air.[7]

The waste storage area consists of six waste pits, two silos, drum storage areas, and scrap piles. The waste pits or settling ponds contain more than 12 million pounds of uranium in a toxic soup generated in 34 years of uranium processing. Private drinking wells south of the Fernald plant are highly contaminated, with some radiation readings as much as 250 times greater than maximum levels set by the Environmental Protection Agency. Readings of uranium concentrations at the Southwest Ohio Water well field are at ten times the level of background radiation (from natural sources), indicating contamination from the center.

HANFORD

The Hanford site is located in semiarid southeastern Washington, 20 miles from the Yakima Indian Reservation. The Yakima River flows along the southern boundary, joining the Columbia River, which flows through the northern part and along the eastern boundary of the site. The site size is 570 square miles. Richland, Pasco, and Kennewick (the Tri-Cities area, population 140,000) are situated nearby, downstream on the Columbia River; Portland, Oregon (population 360,000), is about 230 miles downstream.

Since 1943, the Hanford facility has produced and extracted plutonium for nuclear warheads. As a result, an enormous amount of radioactive waste has been generated. Billions of gallons of contaminated water have been poured into seepage beds, producing an underground plume of

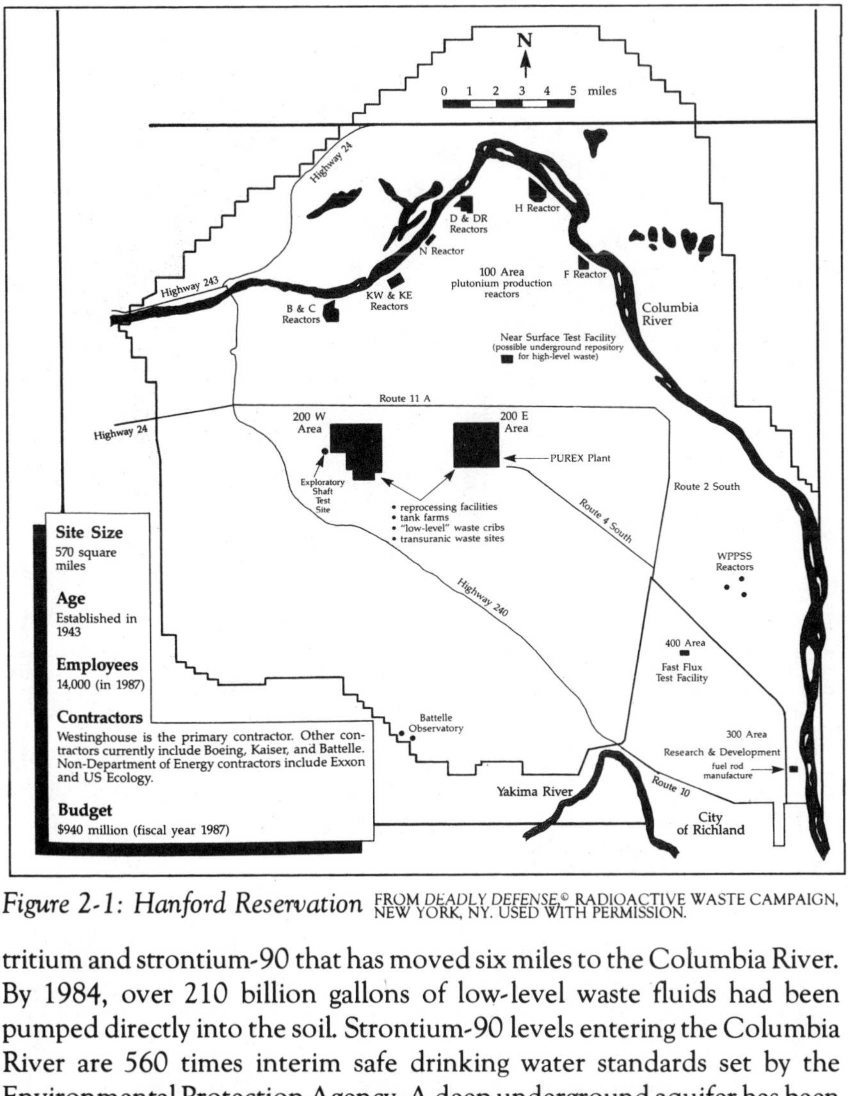

Figure 2-1: Hanford Reservation FROM *DEADLY DEFENSE*,© RADIOACTIVE WASTE CAMPAIGN, NEW YORK, NY. USED WITH PERMISSION.

tritium and strontium-90 that has moved six miles to the Columbia River. By 1984, over 210 billion gallons of low-level waste fluids had been pumped directly into the soil. Strontium-90 levels entering the Columbia River are 560 times interim safe drinking water standards set by the Environmental Protection Agency. A deep underground aquifer has been contaminated with iodine-129. More than 8 million cubic feet of high-level waste, in liquid or slurry form, sit in underground tanks, from which more than 500,000 gallons of high-level waste have leaked. Over 18 million cubic feet of low-level waste and 3.9 million cubic feet of plutonium-contaminated waste are buried at Hanford. Although the plutonium production reactor was shut down in 1987, the Purex reprocessing plant continued to reprocess the irradiated fuel backlog and produce additional waste until December 1989, when an accident forced a shutdown, expected by DOE to be temporary.

IDAHO NATIONAL ENGINEERING LABORATORY

The Idaho National Engineering Laboratory is located in southeastern Idaho, 22 miles west of Idaho Falls. The site is huge, 870 square miles—three-quarters of the size of Rhode Island. The laboratory sits on the Snake River plain, the recharge area for the Snake River aquifer. At the laboratory, enriched uranium is separated from the irradiated naval and experimental reactor fuel that is shipped there by rail from ocean ports. Sixteen billion gallons of waste water, containing 70,000 curies of radioactivity, have been dumped into the Snake River aquifer, which discharges into the Snake River, which in turn flows into the Columbia River near Hanford. This irresponsible practice was temporarily suspended in 1984. The radioactive landfill itself is 140 acres, containing plutonium-contaminated waste from Rocky Flats. More than 350,000 cubic feet of high-level waste, representing 61 million curies, sits in high-level waste tanks as a solid. More than 5 million cubic feet of low-level waste, and 2 million cubic feet of plutonium-contaminated waste are buried in landfills. Another 2.2 million cubic feet of plutonium-contaminated waste from the Rocky Flats Plant are stored at the lab, awaiting transport to the WIPP underground storage facility in New Mexico.

LIVERMORE

The Lawrence Livermore National Laboratory and Sandia National Laboratory at Livermore are located in the Livermore Valley three miles east of Livermore (population 50,000), 40 miles east of San Francisco (population 700,000). The entire Bay Area population of 6 million is within 20 to 75 miles. The town of Tracy (population 25,000) is only six miles from Site 300, where high explosives are tested. The site sizes are 826 acres at Livermore, 221 acres at Sandia, and 7,000 acres for Site 300. In 1942, the laboratory site was a navy base. In 1951, the University of California began using some of the facilities for nuclear weapons research. In 1952, the radiation lab was established. In 1953, Site 300 was established, and high-explosive experimentation began in 1955.

The primary mission of Lawrence Livermore National Laboratory is the design of new nuclear weapons, including "Star Wars" components. Sandia National Laboratories at Livermore develops new warhead components and puts the entire warhead together for testing at the Nevada Test Site. The labs sit above an important aquifer for the Livermore Valley.

Site 300 is leaking tritium into groundwater beyond the site boundary and above state limits. Uranium and tritium are vaporized during open air testing. About 320,000 cubic feet of low-level waste, including 81,000 pounds of uranium, are buried at the lab.

LOS ALAMOS

Los Alamos National Laboratory is located 25 miles northwest of Santa Fe and 60 miles north of Albuquerque, New Mexico. There are 170,000 inhabitants within a 50-mile radius. The lab is on a 43-square-mile site on a plateau between a mountain range to the west and a valley to the east. The land was acquired November 25, 1942.

Los Alamos continues to be the major design center for nuclear warheads. The lab is also now engaged in a large-scale blending operation, combining fuel-grade plutonium from Hanford with super weapons-grade plutonium from the Savannah River Plant. This blend is then transported to Rocky Flats.

During flash floods, plutonium and other radionuclides are washed down the canyons. Movement of radionuclides also is facilitated by wind erosion and burrowing animals. Large-scale transport of plutonium presents a potential threat through accidents and dispersion of plutonium. Over 6.7 million cubic feet of low-level waste (980,000 curies at burial), and 500,000 cubic feet of plutonium-contaminated waste are buried in landfills.

MOUND

The 306-acre Mound Laboratory in Miamisburg, Ohio, 10 miles southwest of Dayton, sits atop the Great Miami aquifer and adjacent to the Great Miami River. About 3.4 million people live within 50 miles of the plant, including residents of Cincinnati and Dayton. Mound produces detonators for activating explosives in nuclear warheads, recovers tritium from weapons components, and constructs plutonium heat generators for satellites. Tritium work began at Mound in 1954.

Accidents and radioactive leakage at Mound have contaminated workers, vast areas of the site, and adjacent public ponds and park land. Tritium and plutonium are contained in sediments of the Old Erie Canal.

NEVADA TEST SITE

The Nevada Test Site, 65 miles northwest of Las Vegas and roughly 20 miles east of the California border, covers 1,350 square miles, two-thirds the size of the state of Delaware. Nuclear weapons are assembled and tested at the site. The first test was conducted January 27, 1951. Through 1988, about 100 atmospheric tests and far more than 600 underground tests have taken place there.[8] The site also serves as a waste disposal facility for other nuclear weapons facilities.

Underground tests have released enormous quantities of radioactive material at or near a major underground aquifer. The soil is heavily contaminated with plutonium and other radionuclides from aboveground tests and breakthrough of underground tests. Almost 8 million cubic feet of low-level waste, including 3.8 million pounds of uranium, are buried in a landfill.

OAK RIDGE RESERVATION

The Oak Ridge Reservation is in eastern Tennessee, about 15 miles west of Knoxville (population 183,000). About 680,000 people live within a 50-mile radius of the 37,000-acre site. The reservation includes three large facilities: the Gaseous Diffusion Plant, the Oak Ridge National Laboratory, and the Y-12 Plant.

The **Gaseous Diffusion Plant** began operations in 1945, producing the enriched uranium used in the Hiroshima bomb. It has been shut down since 1985, and now serves as a waste dump for the Y-12 Plant and the Oak Ridge National Laboratory. More than 2.7 million cubic feet of low-level wastes are buried at the Gaseous Diffusion Plant. During its operation, 36,000 pounds of uranium were released to water and 23,000 pounds to air. Holding ponds and burial grounds contain up to 124,000 pounds of uranium. Large quantities of water-soluble technetium-99 are contained within the piping.

The **Oak Ridge National Laboratory** was originally selected as the location for the world's first uranium chain reactor with production potential. Today the lab functions as a nuclear energy research center and as the Energy Department's center for isotope production and distribution. Large quantities of strontium-90 and cesium have migrated from the six deteriorating radioactive landfills. Over 7 million cubic feet of low-level waste, containing 670,000 curies, are buried at the lab. An aquifer surfaces in the center of burial ground #4. About 940,000 curies were discharged into pits and trenches through 1965, when the practice was halted, and 1.7 million curies of liquid intermediate waste were pumped into underground shale formations. Sediments in White Oak Lake, behind White Oak Dam, contain 600 curies of cesium-137.

The **Y-12 Plant** is located in the Bear Creek Valley, at the northeast end of the Oak Ridge Reservation, and covers 3,400 acres. The major function of the plant is to produce machined components in the internal nuclear weapons assembly. The Y-12 Plant produces the entire lithium-deuteride second stage of the nuclear warhead, and the uranium, beryllium, and other parts of the first stage. In addition, the Y-12 Plant produces highly enriched uranium fuel for the plutonium production reactors at the Savannah River Plant. More than 50 million pounds of uranium are contained within 3.9 million cubic feet of low-level waste in intimate contact with groundwater. Bear Creek and the North Fork of Poplar Creek are contaminated with uranium.

PADUCAH

The Paducah Gaseous Diffusion Plant is 11 miles west of Paducah (population 35,000) and about 3 miles south of the Ohio River and the Illinois border in western Kentucky. The plant takes up 750 acres and is surrounded by a buffer zone. The Paducah Plant enriches uranium for use in the Portsmouth Gaseous Diffusion Plant and the Hanford plutonium production reactor (via Fernald). Since 1952, over 7 million pounds of uranium have been buried on-site; 60,000 pounds have been released into local creeks and 130,000 pounds into the atmosphere. Over 270,000 cubic feet of low-level waste are buried at Paducah.

PANTEX

The Pantex Plant occupies 13,000 acres in the Texas Panhandle, 17 miles northeast of downtown Amarillo (popluation 150,000). The plant was originally built in 1942 for conventional explosives. Nuclear weapons production began in 1951. The principal operation performed at Pantex is the assembly and disassembly of nuclear weapons. High explosives are fabricated and combined with plutonium triggers, second stage components, and other components. Nuclear warheads are then shipped to military installations across the country. The principal danger arises from the transportation of weapons components, toxic chemicals, and nuclear warheads into and out of the Pantex facility. In September 1967, a tornado or hailstorm is known to have heavily damaged the plant and numerous nuclear warheads. The exact damage to warheads remains a classified secret.

PORTSMOUTH

The Portsmouth Gaseous Diffusion Plant is located on 3,500 acres in south central Ohio, 2 miles south of Piketon, 16 miles north of Portsmouth, and 20 miles north of the Ohio-Kentucky border and the Ohio River. Operations began in 1954. The function of the plant is to enrich uranium up to 4 percent uranium-235 for use in commercial nuclear power reactors, and to 93 percent for use in naval, experimental, and Savannah River Plant plutonium production reactors, and in nuclear warheads. Since 1954, more than 11,000 pounds of uranium have been buried on-site, 17,000 pounds have been dumped into local streams, and 23,000 pounds emitted into the atmosphere. More than 350,000 cubic feet of solid low-level waste are buried at Portsmouth.

ROCKY FLATS

The Rocky Flats Plant is located on 6,550 acres at Golden, Colorado, 16 miles northwest of downtown Denver. Almost 2 million people live within 30 miles of the plant, whose primary function in the weapons complex is to produce plutonium triggers for nuclear warheads. Operations began in 1952. With the cessation of plutonium production at

Hanford's N-reactor and the Savannah River Plant reactors since the Chernobyl accident, recycling of nuclear warheads at the Rocky Flats Plant has become the major means of providing plutonium for new warheads. Fires at the plant have distributed a large amount of plutonium into the air, soil, and water supplies. As a result of a contamination incident in October 1988, the plutonium "room" was shut down for several months. In the past, plutonium-contaminated waste has been shipped to the Idaho National Engineering Laboratory. These shipments were halted by the governor of Idaho for a time during 1988, and since September 1989, until the WIPP site is opened in New Mexico. Following an investigation by the FBI, and an argument with federal officials over possible criminal violations, Rockwell Industries announced in September 1989 it was giving up management of the Rocky Flats facility.

SANDIA

Sandia National Laboratories is located immediately south of Albuquerque (population 330,000), within the boundaries of the Kirtland Air Force Base. In Albuquerque, Sandia controls almost 40,000 acres. The Isleta Indian Pueblo, which borders the New Mexico site, has a population of about 2,000. At the Tonopah Test Range in Nevada, Sandia uses almost 370,000 acres. Sandia designs and tests the arming, fusing, and firing systems used in nuclear warheads and bombs, and does research on X ray lasers for the proposed "Star Wars" system. The site has been in operation since 1945. Sandia coordinates and monitors the movement of nuclear materials within the weapons complex. The greatest dangers are movement of radionuclides during flash floods and transportation accidents involving high explosives.

SAVANNAH RIVER PLANT

The Savannah River Plant is located next to the Savannah River, 13 miles southeast of Aiken, South Carolina (population 15,000), and 20 miles south of Augusta, Georgia (population 50,000), on a site of 300 square miles. The plant began operations in 1952. The Savannah River Plant produces nuclear materials, primarily plutonium-239 and tritium, for nuclear weapons. The plant contains three nuclear reactors—currently shut down—and two reprocessing plants for chemical separation of plutonium, tritium, and other nuclear materials.

The Savannah River site contains enormous quantities of nuclear waste in landfills, seepage basins, and 51 underground tanks. About 4.5 million cubic feet of high-level waste, with activity of 790 million curies, are stored in underground tanks. Almost 18 million cubic feet of low-level waste and 160,000 cubic feet of plutonium-contaminated waste are buried at the Savannah River Plant. In addition, over 190,000 cubic feet of plutonium-contaminated waste, containing 530,000 curies, await transport to the WIPP facility in New Mexico. About 30 million gallons of radioactive effluent are discharged annually into seepage basins, at 43,000 times the levels for safe drinking water set by the Environmental Protection Agency. Substantial leaching of burial grounds and surface impoundments is causing severe contamination of the shallow aquifers and groundwater, thus endangering the Tuscaloosa aquifer, which is already contaminated by toxic chemicals from the plant. In 1985, monitoring wells showed increasing spread of contamination into surface streams, underground aquifers, and off-site.

It would be comforting to be able to say that these descriptions refer to past practice, and that the Department of Energy has now become open, honest, responsible, and a good neighbor. But such, unfortunately, is not the case. Despite the inevitable and clearly documented consequences of disposing of radioactive materials into landfills, pits, and cribs, the practice continues at Department of Energy facilities. Many of the practices begun under wartime emergency conditions in World War II have continued to this day. Insulated by the national security cover, and particularly by the congressional Armed Services committees, the nuclear weapons complex has successfully resisted the changes that have taken place in other agencies and in the commercial sector over the past 20 years in response to environmental concerns.

An encouraging sign is the awakening environmental ethic of the American people. Beginning with the National Environmental Policy Act, and continuing with the Resource Conservation and Recovery Act, citizens have been able to pry off the cover of secrecy and view Department of Energy practices in their full light. The public is now calling for action and for the same adherence to environmental standards as is required of the commercial sector.

It is clear that the cleanup cost will be high. Assuming an effective cleanup were possible, in July 1988, the department projected the costs to run between $40 and $110 billion. Another $35 billion was estimated for consolidation and modernization of the aging weapons complex. The General Accounting Office raised the total estimate for the former and the latter to $175 billion[9]. This is to be contrasted with the annual nuclear weapons budget of $8 billion, which includes up to $1 billion for health and safety projects, among them waste management. At present budgetary levels, it would take over 100 years to clean up past practices. More recently, the Department of Energy's "2010 Report"[10], approved by President Bush's Office of Management and Budget, lowered the estimated cleanup costs to $29 billion. Diverse assumptions account for this wide range in expected costs. Exhumation of the waste pits at Fernald could cost $2 billion, whereas replacing and contouring the cover over the waste pits could cost $50 million. Solidification and disposal of high-level waste in older single shell tanks at Hanford alone could cost up to $45 billion.

The nuclear weapons complex is clearly in crisis. While cleaning up shoddy past practices, the Department of Energy must also consolidate and replace its aging plant. Difficult decisions lie ahead for Congress as it begins to factor the cost of waste management into the cost of nuclear weapons production. Congress and the Bush Administration may want to cut corners and get back to "business as usual." Environmental groups will have to keep their feet to the fire, to protect the public health and safety. And if the cost is too high, maybe fewer bombs and more disarmament will result. At an estimated cost of over $6 million for each nuclear weapon in our nuclear arsenal, TNT is beginning to look cheap in comparison.

NOTES

1. *Integrated Data Base for 1987*, Department of Energy, DOE/RW-0006, Washington, D.C., September 1987.
2. *Deadly Defense*, Radioactive Waste Campaign, New York City, June 1988. Much of the information about nuclear weapons facilities was obtained from Department of Energy sources and collected in the Campaign's book.
3. *Environmental Monitoring Report for 1987*, Department of Energy, Savannah River Plant, 1988.

4. *The Management of Radioactive Waste at the Oak Ridge National Laboratory: A Technical Review,* National Academy of Sciences, National Academy Press, Washington, D.C., 1984.
5. See column by Frank Munger, *Knoxville News-Sentinel,* Knoxville, Tenn., March 1986.
6. NRC Information Notice, Office of Nuclear Material Safety and Safeguards, Nuclear Regulatory Commission, July 1988.
7. "Uranium Releases at Fernald, Radiation Doses to Nearby Residents," Marvin Resnikoff, Radioactive Waste Campaign, June 7, 1988.
8. *Nuclear Weapons Databook, Vol. 2,* "U.S. Nuclear Warhead Production," T. Cochran et al., Natural Resources Defense Council, Ballinger Publishing Co., Cambridge, Mass., 1987.
9. "Dealing with Major Problem Areas in the Nuclear Defense Complex Expected to Cost Over $100 Billion," General Accounting Office, GAO/RCED-88-137, July 13, 1988.
10. U.S. Department of Energy, Nuclear Weapons Complex Modernization Report, "Report to the Congress by the President," Washington, D.C., December 1988.

3

The Bomb Factories: Out of Compliance and Out of Control

DAN W. REICHER

S. JACOB SCHERR

THE U.S. DEPARTMENT OF ENERGY (DOE) FACES A CRISIS OF UNPRECEDENTED proportions at its vast nuclear weapons production complex. After decades of mismanagement and neglect and with ongoing revelations of safety and environmental problems at plants around the country, it is now clear that the continued operation of DOE's aged weapons facilities represents one of the most serious safety and environmental threats facing our nation. The beleaguered weapons complex, long shielded from public scrutiny and outside regulation, is riddled with scores of dangerous production plants and hundreds of leaking toxic and radioactive waste dumps. According to a top DOE official, the prevailing attitude toward safety among some department managers is similar to the one that existed at NASA prior to the Challenger disaster. Among the multitude of problems plaguing the complex are the following:

- Many significant accidents, including the melting of nuclear fuel, have occurred over the past 30 years at DOE reactors at the Savannah River Plant in South Carolina, according to DOE documents made public in 1988. Physicists have described these accidents, many of which were kept secret for more than three decades, as among the most severe ever documented at a U.S. nuclear plant. The SRP reactors have been shut down pending the resolution of safety problems.

- At the Hanford Nuclear Reservation, in Washington state, more than 200 billion gallons of liquid wastes—enough to cover Manhattan to a depth of 40 feet—have been discharged to unlined ponds and trenches, contaminating over 100 square miles of groundwater with radioactive and chemical wastes. Tanks filled with highly radioactive liquid wastes have leaked more than 200 kilograms of toxic plutonium into the ground—enough for nearly 50 Nagasaki-sized bombs. In 1986, springs discharging to the Columbia River from the Hanford site had concentrations of radioactive materials hundreds of times higher than federal standards.
- DOE knowingly released thousands of tons of uranium waste from the Feed Materials Production Center near Cincinnati, Ohio, exposing workers and nearby residents to radiation. Uranium from the plant leaked into nearby drinking water wells; neighbors used the wells for almost two years, until DOE finally notified them about the contamination. DOE agreed to pay over $70 million in settlement of a case brought by 1,400 Ohio residents who claimed that the plant had diminished the value of their property and harmed their psychological well-being.
- At the 890-square-mile Idaho National Engineering Laboratory, billions of gallons of waste have been injected deep into the ground. Contaminants, including carcinogenic solvents, have reached the Snake River aquifer, a vital water source in the northwestern United States. Plutonium escaping from waste sites at the facility is migrating toward the aquifer.
- In 1989, the U.S. Federal Bureau of Investigation launched a major investigation of alleged criminal violations of federal environmental laws at the Rocky Flats Plant in Colorado. In 1988, DOE was forced to close a major plutonium-processing facility at Rocky Flats after workers were accidentally contaminated with plutonium. In 1957 and 1969, major plutonium fires occurred at Rocky Flats, which is located 16 miles from Denver. Plutonium, an extremely carcinogenic substance, was subsequently found downwind of the plant in significant concentrations.

The recent crisis at the weapons production complex has finally forced Congress and the public to come to grips with the true cost of nuclear weapons production. Estimates released in July 1988 by DOE and the

General Accounting Office, a research arm of Congress, indicate that $100 billion or more could be required for environmental cleanup and waste disposal at DOE facilities, more than the United States spent to put a man on the moon. This is on top of the $250 billion that the nation has already spent to date on the production of nuclear warheads. This disturbing legacy of the nuclear arms race reflects, in part, the absence of full and independent oversight of the federal government's nuclear weapons production activities and DOE's resistance to compliance with federal environmental laws.

In the Atomic Energy Act (AEA) of 1954, Congress subjected civilian nuclear facilities to licensing by the Atomic Energy Commission, including public hearings and disclosure of information. However, in the Cold War atmosphere of the early 1950s, congressional concerns about national security precluded public oversight of the government-owned nuclear weapons production complex. This regulatory vacuum was maintained in the Energy Reorganization Act of 1974, which split the Atomic Energy Commission into the Nuclear Regulatory Commission (NRC) and the Energy Research and Development Administration, which later became DOE. The NRC was given the task of regulating the safety of commercial nuclear facilities and the handling of radioactive materials. The commission's authority, however, was not extended to the DOE weapons plants. The department was left to police the safety of its own facilities without any outside oversight.

During the 1970s, Congress created the Environmental Protection Agency (EPA) and enacted a series of environmental protection statutes that applied to both commercial and government facilities. Under some of the statutes, the president can exempt federal facilities from compliance when it is in the paramount interest of the U.S., but this national security exemption has rarely been invoked.

DOE has nonetheless vigorously resisted the full application of these critical environmental laws at its weapons facilities. The department has viewed compliance as an interference with its primary mission of nuclear weapons production and an expensive diversion of resources. This attitude is exacerbated by the tradition of secrecy surrounding the nuclear weapons complex and DOE's deep-seated resistance to change. One top Environmental Protection Agency official characterized the department's approach toward environmental laws as, "Look, Buster, don't bug me with your crap about permits. I'm building atomic weapons."

The DOE's disturbing lack of accountability at the weapons plants and the serious consequences that result are well illustrated by controversies at four major facilities: DOE's refusal to prepare an Environmental Impact Statement (EIS) on the renovation of the L Reactor at the Savannah River Plant (SRP) in South Carolina; violations of federal hazardous waste laws at the Y-12 Plant at Oak Ridge, Tennessee; the department's proposal to restart the N Reactor at Hanford, Washington, in the wake of the Chernobyl disaster; and DOE's efforts to dispose of radioactive wastes in a potentially unsound geologic repository near Carlsbad, New Mexico.

THE L REACTOR RENOVATION

The L Reactor is one of five weapons materials production reactors built in the early 1950s at the Savannah River Plant along the border between South Carolina and Georgia. The reactor, which lacks the radiation containment structures and cooling towers incorporated in modern power reactors, was shut down and placed on standby status in 1968 due to a decreased demand for nuclear weapons materials. In 1982, DOE's plans to overhaul and restart the L Reactor without an Environmental Impact Statement (EIS) created an uproar.

An EIS must be prepared for all "major Federal actions significantly affecting the quality of the human environment," according to the National Environmental Policy Act of 1969 (NEPA). By the end of the 1970s, DOE and its predecessor agencies had integrated the EIS requirement into its decision-making processes and had prepared more than 30 such statements on activities at a number of its nuclear weapons production facilities.

With the start of the Reagan administration in 1981, a change in priorities set the stage for the L Reactor controversy. The administration decided that the production of nuclear weapons materials was to be stepped up and signaled that environmental protection was a secondary concern. As William A. Vaughan, DOE assistant secretary for Environmental Protection, Safety, and Emergency Preparedness during much of President Reagan's first term, told the *New York Times* in November, 1988, "There was a military culture throughout the agency, a bunker mentality. People saw their role as meeting the requirements for defense. Other things, of necessity, came secondary, including environmental, safety, and health programs. Every penny that went to safety programs was a penny taken from manufacturing nuclear warheads."

At the same time, administration supporters on the House and Senate

Armed Services committees introduced legislation to limit sharply the requirement that EIS's be prepared on DOE defense activities. Their view was that previous EIS's on the department's continuing activities were a source of public agitation and a waste of money. Although this legislation was reduced to a requirement that DOE notify the Armed Services committees before initiating an EIS, the lack of support in the committees for NEPA's stringent application to nuclear weapons materials production was made clear.

In 1981, DOE initiated over $200 million worth of renovations for the L Reactor, which had sat idle for 13 years with virtually no maintenance. It should have been obvious that an EIS was required. The restoration was clearly a major federal action, the L Reactor had never been the subject of an EIS, and since the passage of NEPA not a single nuclear reactor of this size had come into operation in the United States without the preparation of an EIS. As with other reactors, the operation of the L Reactor would have considerable environmental impact, including the routine and accidental release of radioactive materials and the generation of nuclear wastes. Also, lacking cooling towers, the L Reactor would discharge over 175,000 gallons per minute of scalding water into a tributary of the Savannah River in violation of water quality standards. DOE's own studies showed that this discharge would destroy about 1,000 acres of productive wetlands and would flush out radioactive cesium long trapped in sediment in the Savannah River. There was also concern about the impact of discharges on endangered wood storks living in the area. DOE argued that these effects were "insignificant" when compared with the effects of L Reactor operations before 1968, and thus decided not to prepare an EIS.

DOE's plans to restart the L Reactor were met by protests, particularly from officials and citizens living downstream and downwind of SRP in the rapidly growing coastal areas of South Carolina and Georgia. In November 1982, a coalition of citizen groups and individuals filed suit to force DOE to prepare an EIS. Even though SRP, with 15,000 workers, is one of the largest employers in South Carolina, the state joined the legal action. DOE was particularly disturbed by South Carolina's participation in the case and later attempted unsuccessfully to negotiate a separate settlement.

Senator Strom Thurmond (R-SC) also tried to quell the dispute by seeking the opinion of the Environmental Protection Agency (EPA), which has a formal role in reviewing and commenting on the environmental effects of other federal agency activities. At a Senate Armed

Services Committee hearing on the L Reactor in Aiken, South Carolina, Senator Thurmond released a letter from EPA Administrator Anne Gorsuch that fully supported the DOE decision not to undertake a full NEPA review. Less than two months later, on the eve of the administrator's resignation, EPA documents released under the Freedom of Information Act revealed that Gorsuch had ignored her own staff's technical analyses, which had concluded that the start-up of the L Reactor would indeed have significant environmental effects.

The affair was resolved in an unusual coincidence of court and congressional action. On July 15, 1983, the U.S. District Court for the District of Columbia ruled that an EIS had to be prepared on the L Reactor. The judge held that the effects of the proposed reactor start-up were significant when measured against environmental conditions at that moment and not, as DOE contended, those associated with reactor operations years earlier. The same day, an amendment to the Defense Appropriations Bill was signed into law delaying the start-up of the L Reactor until the EIS was completed.

The EIS process, which took ten months, resulted in tens of millions of dollars worth of environmental protection measures, including the construction of a cooling lake to assure compliance with water quality standards and the start of efforts to clean up contaminated groundwater at SRP. However, DOE refused to make further improvements to the reactor's safety systems, which it claimed were adequate to protect public health and the environment. The start-up of the L Reactor was delayed until October 31, 1985, some two years beyond the date by which DOE had contended that operation of the reactor was essential for national security.

Moreover, DOE's claims in the EIS about the safety of the SRP reactors have subsequently been shown to be false. For example, the department stated in the EIS that "no significant reactor accidents have occurred at the SRP in its 30 years of operation" and that "fuel melting has never occurred in the SRP reactors." But recently disclosed DOE documents leave no doubt that significant accidents have occurred at the SRP reactors during the last three decades, including fuel melting.

The L Reactor as well as the two other operable SRP reactors are currently shut down, pending the resolution of safety problems. In the wake of litigation brought by environmental groups, the department has agreed to prepare a comprehensive EIS before starting up any of the plants.

Y-12 PLANT HAZARDOUS WASTES

In 1983, DOE admitted after years of cover-up that the Y-12 weapons materials plant at Oak Ridge, Tennessee, had dumped or spilled into nearby streams some 2.4 million pounds of mercury from 1950 to 1963. Fish consumed by local residents were contaminated with mercury significantly above federal limits. In addition, four waste disposal ponds at Y-12 were also revealed to be leaking 4.7 million gallons of metal plating wastes, acids, and solvents into groundwater each year. One expert described DOE waste management practices at Y-12 as "state-of-the-art 1948."

Waste treatment, storage, handling, and disposal in the DOE nuclear weapons complex are subject to the Resource Conservation and Recovery Act (RCRA) of 1976, which establishes a "cradle-to-grave" management system for toxic chemical wastes with regulatory authority shared by EPA and state agencies. However, when RCRA was enacted, DOE claimed a complete exemption for all of its facilities. In 1983, after years of foot-dragging by federal and state agencies, the Legal Environmental Assistance Foundation and the Natural Resources Defense Council filed suit to compel DOE to comply with RCRA at Y-12.

DOE argued that application of RCRA to the DOE nuclear weapons complex was inconsistent with its weapons production responsibility under the Atomic Energy Act because it would subject the facilities to outside regulation by EPA, states, and localities. According to the department, such outside control would jeopardize national security. In April 1984, the federal district court in Tennessee rejected DOE's assertions and ruled that the department must obey RCRA.

DOE did not appeal the Y-12 decision but continued to resist implementation of RCRA at its facilities. DOE quickly claimed an exemption for all its chemical wastes that are mixed with radioactive materials, subject to the Atomic Energy Act. These "mixed wastes" represent a substantial proportion of the department's wastes.

Under pressure from citizen groups and members of Congress, DOE subsequently agreed with EPA that mixed wastes are subject to RCRA. However, in November 1985, the department attempted another ploy to limit RCRA's application to these wastes. DOE proposed a rule that would redefine much of the mixed wastes as "byproduct material," a category of radioactive material under the AEA that is explicitly exempt from RCRA coverage. DOE was particularly intent on preventing application of RCRA to its leaking tanks of highly radioactive wastes at the Hanford Nuclear Reservation and the Savannah River Plant.

States and citizen groups criticized the proposed byproduct rule as yet another attempt by DOE to avoid its environmental responsibilities, and 70 members of Congress signed a letter calling upon DOE Secretary John Herrington to abandon the rule. In May 1987, DOE finally backed down and agreed that all of its mixed wastes are subject to RCRA. However, in the same document, DOE again raised the possibility that inconsistencies between RCRA and the department's obligations under the Atomic Energy Act might preclude full application of RCRA. Faced with a cleanup bill of over $100 billion, DOE may argue that hazardous waste management must yield to the department's weapons production responsibilities.

N REACTOR RESTART

The fallout from the Chernobyl accident included heightened concern about the safety of the N Reactor, a 25-year-old plutonium production reactor located on the Columbia River at the DOE Hanford Reservation in Washington state. Like the ill-fated Soviet reactor, the N Reactor is graphite-moderated and light water–cooled. Neither reactor is housed in a containment structure of the type used on most U.S. commercial reactors to prevent the escape of radiation in an accident. The N Reactor instead relies on a "confinement system," which is a series of filters housed in a standard industrial building surrounding the facility.

Following the Soviet disaster, the Energy Department continued to operate the N Reactor. After considerable public and congressional pressure, however, DOE appointed a panel of six outside experts and requested the National Academy of Sciences to review the safety of its large reactors.

The panel, chaired by Louis Roddis, former president of Consolidated Edison of New York, submitted its reports to DOE Secretary Herrington in early October, 1986. The reports were not made public, however, until after the November elections of that year, which included a hotly contested Senate race in Washington that focused heavily on Hanford. In an unprecedented move, Chairman Roddis and Professor Harold Lewis of the University of California called for permanent shutdown of the N Reactor unless national security interests outweighed the safety risks posed by the facility. Moreover, all six panelists agreed on three important points in their separate reports. First, in the event of an accident a hydrogen explosion could occur at the N Reactor (the buildup of hydrogen was a significant concern in both the Chernobyl and Three Mile Island accidents). Second, the N Reactor does not provide a level of safety comparable to commercial reactors. Third, the N Reactor is not subject to adequate outside safety review.

The panel members noted that as early as 1958 the Atomic Energy Commission's Advisory Committee on Reactor Safeguards had concluded that it was "essential" that the N Reactor be equipped with a containment structure. This advice was not heeded when the N Reactor was built, leading the advisory committee to conclude in 1966 that the "radiological consequences . . . of a severe reactor accident at the N Reactor . . . would exceed the dose guidelines . . . for licensed power reactors." Twenty years later, Chairman Roddis found this situation "unchanged."

In January 1987, after the release of the Roddis Panel reports, DOE shut down the N Reactor temporarily and began over $100 million in safety modifications. But the modifications did not address the reactor's fundamental flaw: the lack of containment. Following the shutdown, the National Academy of Sciences (NAS) also released a report critical of the safety of the N Reactor, as well as the SRP reactors. (A preliminary report from the NAS six months earlier had resulted in the SRP reactors being reduced to half power.) In a press conference following the release of the NAS Report, DOE Under Secretary Joseph Salgado admitted for the first time that the DOE reactors did not provide a margin of safety comparable to commercial reactors.

DOE's intent to restart the N Reactor in March 1988 spurred intense debate in Congress, especially among representatives from Washington and Oregon. The Senate Armed Services and Appropriations committees went so far as to call for the moth-balling of the plant. Congress reached a compromise which required that the Secretary of Energy certify the safety of the reactor. Congress also indicated that if there was no "timely restart" of the facility, then the reactor was to remain shut down indefinitely and its operating funds devoted to development of new plutonium production and enrichment facilities. Additionally, 12 citizen groups and 6 members of Congress filed suit to prevent restart of the N Reactor pending preparation of a full Environmental Impact Statement.

DOE pushed forward with its plans to restart the facility early in 1988, arguing that it was critical to national security. According to one press account, the department went so far as to amend the secret document that lays out the nation's nuclear weapons materials requirements—the Nuclear Weapons Stockpile Memorandum—to underscore the need for the restart.

It was thus a great shock when the department announced in February 1988 that it had decided not to restart the facility, but instead to place it in "standby status" for emergency use. Testifying before a congressional subcommittee about the department's decision, Secretary Herrington sounded much like the activists he had been resisting for years: "We're awash in plutonium. We have more than we need."

In October 1988, it was revealed that the department was analyzing the feasibility of using the N Reactor to produce tritium, another ingredient in nuclear weapons. However, any attempt to restart the N Reactor for this purpose would probably result in lengthy technical, legal, and legislative battles.

RADIOACTIVE WASTE DISPOSAL IN NEW MEXICO

In 1979, the U.S. Congress approved DOE plans to construct the Waste Isolation Pilot Plant (WIPP) in an underground salt formation near Carlsbad, New Mexico. WIPP, which would be the world's first geologic repository expressly for the isolation of nuclear waste, is designed to hold almost one million drums of plutonium-contaminated materials from DOE weapons programs.

Over the years, the project has faced a variety of challenges. In 1980, President Carter sought to cancel WIPP, in part because Congress exempted it from Nuclear Regulatory Commission licensing and oversight. Twice during the Reagan administration, the White House Office of Management and Budget tried to eliminate WIPP because of its near billion-dollar cost. Then, in 1986, the U.S. General Accounting Office revealed that WIPP will hold only a small fraction of DOE defense wastes and that the facility will not be available for the wastes most in need of clean-up—solid wastes buried in shallow pits or liquid wastes drained into the ground.

WIPP has survived all of these challenges, but as of late 1990, DOE had not resolved an even more daunting technical hurdle. An independent group of scientists from the University of New Mexico revealed in 1987 that water was seeping into what were supposed to be dry underground chambers. According to the scientists, this water could mix with radioactive wastes to produce a "radioactive slurry." Migration of the slurry could contaminate nearby water supplies such as the Pecos River, which flows south through New Mexico and Texas. Moreover, in June 1988, the panel released an unpublished DOE report identifying a large pressurized body of water under the WIPP site that, if released, could carry radioactivity upward to the surface environment.

These problems raise significant questions about the ability of WIPP to contain radioactive wastes safely over the hundreds of thousands of years they will remain dangerous. Nevertheless, DOE sought to open WIPP in October 1988, and to do so attempted to skirt critical Environmental Protection Agency standards.

The EPA issued standards for protection of public health and the environment from the disposal of radioactive waste in 1985. The standards require DOE to show that repositories like WIPP will not, over thousands of years, release radiation in excess of specified limits. Following a legal challenge by a coalition of environmental organizations and states, the First Circuit Court of Appeals in Boston determined in 1987 that the standards may not protect groundwater adequately and ordered EPA to rewrite them.

But DOE did not want to wait for EPA to reissue the standards before it began emplacing waste at WIPP. DOE claimed in 1988 that it needed to load wastes into WIPP as an "experiment" to determine whether the facility is safe. But independent scientists said that waste emplaced at WIPP may be difficult, if not impossible, to retrieve if problems develop and may impede critical improvements to the facility. According to these scientists, any tests with radioactive wastes should instead be done in a laboratory under more controlled conditions.

At a congressional hearing in September 1988, DOE acknowledged that WIPP faced another serious problem. Department officials testified that 80 to 90 percent of the radioactive wastes to be loaded into WIPP also contain nonradioactive materials subject to regulation under federal hazardous waste laws. But DOE had neither conducted any studies on the ability of WIPP to contain these wastes nor secured any permit for their disposal. Following the hearing, the department announced that it had postponed opening WIPP indefinitely. It remains to be seen whether DOE can resolve the complex safety and environmental problems plaguing the facility.

THE NEED FOR OUTSIDE OVERSIGHT

These controversies demonstrate the critical need for outside oversight of DOE facilities. There is simply no independent check on the safety of the department's operations. Moreover, existing laws and regulations have proven inadequate in controlling chemical and other environmental hazards, especially in the face of the administration's recalcitrance. With the department's cleanup bill rising and a number of new multibillion-dollar production facilities on the horizon, the need for oversight is all the more pressing.

What is required is an ongoing independent oversight body with the expertise to identify problems in the weapons production complex and the authority to see that they are corrected. The Nuclear Regulatory Commission is technically capable of safety oversight, but lacks the authority. Congress has authority over the facilities, but is ill-equipped technically. The courts lack both technical competence and the authority to do much more than force DOE to reconsider its decisions. DOE has formed an advisory panel to oversee the safety of its reactors, but the panel has no authority to compel DOE to implement the panel's recommendations.

Congress has recently considered two basic approaches to outside oversight: expansion of NRC authority to include the weapons complex, and creation of an entirely new oversight body. Under legislation introduced by Representative Ron Wyden (D-Ore) in 1988, the Nuclear Regulatory Commission would review the safety of existing facilities and license and oversee new facilities. At least in theory, the NRC has many of the characteristics necessary for effective oversight: the ability to shut down unsafe facilities, the capacity to issue enforceable sanctions for safety violations, independent on-site inspectors, and formal procedures for full public participation. Admittedly, the agency does not have an inspiring track record in regulating commercial facilities.

Congress, however, has at least temporarily rejected the NRC approach. Under a provision of the 1988 Defense Authorization Act, Congress established a five-member Defense Nuclear Facilities Safety Board. The new board falls far short of what is needed for effective oversight. The original bill, introduced by Senator John Glenn (D-Ohio), would have required DOE to carry out the board's safety recommendations at all of its defense and civilian nuclear facilities unless they were "technically infeasible" or if the president determined that compliance was not in the "paramount interest" of the United States.

Under the legislation as finally adopted, DOE is not bound by the oversight board's recommendations. Certain weapons production facilities are exempt from the board's oversight, including the Pantex plant in Texas, where warheads are assembled, the weapons laboratories, and the Nevada Test Site. The public has no formal role in the board's proceedings other than to comment on its recommendations. Nor is there a provision for permanent on-site inspectors at DOE facilities. And unlike the legislation originally introduced by Senator Glenn, the president can dismiss board members at will.

The Bush administration and Congress should act swiftly to correct the board's serious shortcomings. It is also imperative that Congress impose NRC licensing on the new multibillion-dollar weapons production facilities DOE is seeking to build. The time is long past when we should be constructing and operating unlicensed nuclear reactors and other facilities.

With respect to environmental oversight, the L Reactor and Y-12 controversies demonstrate that having legal authority over DOE may not be sufficient to compel the department's compliance with outside controls. Effective enforcement of environmental laws and regulations has been thwarted not only by DOE's obstinacy, but also by Justice Department policies. According to the Department of Justice, EPA has no authority to bring suit against DOE for violations of federal laws and regulations. The Justice Department also contends that EPA cannot even issue enforceable administrative orders against the department in many circumstances. The EPA has been reduced to bargaining with DOE, a department with far more political clout. This has left enforcement to state agencies and citizen groups, which often have limited resources. Legislation to address these problems is currently pending in Congress. Its enactment should be made a high priority.

As Senator Glenn has said, "We cannot afford to risk permanent harm to our citizens and our land in the name of national security." Although legislative and policy adjustments will help to correct the pressing problems at DOE facilities, the ultimate solution depends upon the department's willingness to agree that protection of human health and the environment is central to its national security mission.

REFERENCES

Finamore, Barbara A., "Regulating Hazardous and Mixed Waste at Department of Energy Nuclear Weapons Facilities: Reversing Decades of Neglect," *Harvard Environmental Law Review,* vol. 9, no. 1, 1985, pp. 83-141.

General Accounting Office, "Environmental Issues at DOE's Nuclear Defense Facilities," GAO/RCED-86-192, September 1986.

General Accounting Office, "Dealing with Problems in the Nuclear Weapons Production Complex Expected to Cost Over $100 Billion," GAO/RCED-88-197BR, July 1988. Also see GAO reports GAO/RCED-88-62, GAO/RCED-88-130, GAO/RCED-87-124, and GAO/RCED-86-132.

Lennett, David J. and Dan W. Reicher, "Statement Regarding the Application of the Resource, Conservation and Recovery Act to the Department of Energy's Planned Waste Management Activities at the Waste Isolation Pilot Plant," before the House of Representatives, Government Operations Committee, Subcommittee on Environment, Energy, and Natural Resources, Sept. 13, 1988.

National Academy of Sciences, National Research Council, "Report on Brine Accumulation in the WIPP Facility," Washington, D.C., March 3, 1988.

National Academy of Sciences, National Research Council, *Safety Issues at the Defense Production Reactors, A Report to the DOE*, National Academy Press, Washington, D.C., 1987.

Reicher, Dan W., "Statement Regarding Federal Facility Compliance with the Comprehensive Environmental Response, Compensation, and Liability Act," before the House of Representatives Committee on Environment and Public Works, Subcommittee on Superfund and Environmental Oversight, Aug. 4, 1988.

Roddis Panel Reports, Reports to Secretary of Energy John Herrington by Miles C. Leverett, H. W. Lewis, Thomas H. Pigford, Louis Roddis, Gerald F. Tate, and E. P. Wilkinson, October 1986.

Schneider, Keith, "Plutonium Leak in Idaho Symptom of Atomic Ills," *New York Times*, April 17, 1988, p. 24.

———."U.S. Delays Start of Plant to Store Nuclear Wastes," *New York Times*, Sept. 14, 1988. p. 1.

———."Accidents at a U.S. Nuclear Plant Were Kept Secret for Up to 31 Years," *New York Times*, October 1, 1988, p.1.

———."2nd Nuclear Plant is Ordered Closed by Energy Dept.," *New York Times*, Oct. 11, 1988. p. 1.

———."U.S., for Decades, Let Uranium Leak at Weapon Plant," *New York Times*, Oct. 15, 1988, p. 1.

———."Defects in Nuclear Arms Industry Minimized in Early Reagan Years," *New York Times*, Nov. 7, 1988, p. 1.

U.S. Department of Energy, *Final Environmental Impact Statement, L-Reactor Operation Savannah River Plant,* vols. 1–3, DOE/EIS-0108, May 1984.

U.S. Department of Energy, "Environment, Safety and Health Report for the Department of Energy Defense Complex," July, 1, 1988.

4

The Military Production Reactors and the Politics of Safety and Security

BENNETT RAMBERG

"We have a serious situation that must be dealt with. Unfortunately, the vulnerabilities of our existing facilities did not get significant attention from the Administration until after the April 1986 Chernobyl accident forced them to take notice."

SENATOR J. BENNETT JOHNSTON, opening statement to Senate Committee on Energy and Natural Resources on the "Department of Energy's Facilities for Defense Materials Production," October 27, 1987.

"Since its founding days, [the Savannah River Plant] operated mainly in secret. It's evident now that concerns over safety, environment and health did not matter at all during those days. All they wanted was production. But people have a right to know what kinds of conditions they are working and living near."

SENATOR JOHN GLENN, commenting on revelations that the Department of Energy had kept secret from the public numerous serious accidents at the Savannah River Plant, *New York Times*, Oct. 4, 1988.

CONSIDER THE FOLLOWING DECISIONS REGARDING THE NUCLEAR WEAPONS material production complexes announced in 1987 and 1988.

- The Department of Energy (DOE) announces that it plans to mothball the Hanford N Reactor, the largest U.S. nuclear production reactor.
- The Department of Energy announces that it will immediately cut power generation at its Savannah River production reactors by one-half; later it suspends all nuclear materials production.
- The Department of Energy announces that it plans to construct one, perhaps two new weapons reactors.

What accounts for these decisions? Why did DOE interrupt and later terminate operations at the N Reactor after more than twenty years of relatively uneventful operation? Why did it cut power generation at Savannah River and later suspend operations there? Why did the Department of Energy after years of procrastination suddenly announce plans in 1988 to build one or two new production reactors?

The answers to these questions may be found in changes in public policy. What induced these changes—and why they took so long in coming—is the subject of this paper. The exploration will focus on six explanatory factors: the public agenda, pressing issues, new players, techno-bureaucratic critics, policy entrepreneurs, and windows of opportunity.

Because policy makers confront a potpourri of items that deserve attention and they have limited time and resources to devote to any one item, they must prioritize. Achieving priority status—getting on the policy agenda—is essential for any new direction in public policy. Once an issue attains this visibility, decision makers and their advisers will examine alternatives. The political process will yield choices that the decision maker will try to implement.[1]

"Pressing issues" are matters that attract the attention of decision makers. How does a matter become pressing? When they assume a decision making role, "new players" bring their own priorities, as in the change of administrations following an election. "Techno-bureaucratic critics" include members of the bureaucracy—in some cases whistle blowers driven by conscience—or institutionalized and ad hoc advisory groups employed by government who challenge conventional wisdom. "Political entrepreneurs" are interest groups outside government who make it their work to lobby policy makers and bureaucracies to reinforce or change policy. Finally, "windows of opportunity" are events that thrust an item onto the public menu, forcing officials to confront issues heretofore neglected.

THE SETTING

Before delving into the politics of nuclear safety at the production reactors, it is important to understand something about the facilities themselves. The nuclear complexes at the Hanford Reservation in the state of Washington and Savannah River in South Carolina trace their origins, respectively, to the dark days of World War II and the onset of the Cold War. In each case there was—and indeed in the minds of some government officials, still is—one pressing agenda item, namely the production of fissile materials to support the nuclear weapons arsenal.

Built in 1963, the 4000 megawatt thermal (MWt) N Reactor at Hanford is the largest of nine reactors built on the site. Until the Department of Energy (DOE) placed it in "cold shutdown" in 1987, it was the only operating reactor at Hanford; the federal government had retired the others years earlier as their useful lives came to an end and demand for weapons grade nuclear material declined.

The N Reactor is a pressurized, heavy-water cooled, graphite-moderated facility. Its safety infrastructure includes 84 horizontal boron-carbide control rods, backup boron-carbide balls, emergency core cooling, and a confinement system with filters to reduce radioactive releases in the event of a serious accident. Until the recent decision to mothball the plant, the Department of Energy expected the reactor would function through the mid to late 1990s, by which time swelling of the graphite moderator and embrittlement of pressure tubes would make operation impractical.[2]

To ensure the production of tritium for thermonuclear weapons and to provide alternative sources of plutonium, President Truman authorized construction of the heavy-water cooled and graphite-moderated Savannah River reactors in 1950. The Atomic Energy Commission built five plants. Three operate today. The K and P Reactors mainly produce tritium; the L Reactor, recommissioned in 1985 following a 17-year hiatus, generates plutonium.[3] Over the years AEC and DOE upgraded the plants to generate a maximum of 2800 MWt. Into these low-temperature and low-pressure reactors, contractors engineered gravity and electrically driven safety and control rods, a liquid injection, and emergency core cooling systems, plus confinement and filtration structures to address accidents.[4]

THE POLITICS OF NUCLEAR SAFETY AT THE PRODUCTION REACTORS BEFORE THREE MILE ISLAND (TMI)

Safety was never the pressing issue on the agenda of energy officials; the pressing issue was the production of nuclear materials for the weapons program. Nonetheless, "acceptable risk" was a bone of contention: some decision makers and advisors wanted significant modifications that would dramatically increase the cost of the plants but make them hypothetically more secure, and others were content with the status quo. Over time, the changing definition of risk complicated matters as experts and the public learned more about nuclear hazards.

Reflecting this dynamic environment, the Atomic Energy Commission, which administered the reactors until 1974, institutionalized an external bureaucratic critiquing mechanism—the Advisory Committee on Reactor Safeguards (ACRS). The AEC intended for the committee to serve as its conscience to place controversial technical and safety matters on the departmental agenda. But just as one's own conscience does not necessarily dictate personal decisions, the ACRS had no responsibility—no institutionalized political clout—to implement recommendations. It served at the pleasure of the AEC.[5]

In 1958, in the spirit of consciousness raising, the ACRS addressed the matter of safety during the planning and engineering phase of the N Reactor. It proposed an elementary criterion: containment must be provided and integrity of the cooling system guaranteed. Of course, the N Reactor was not a civil reactor, although several years after it went into operation it was connected to the northwest electric power grid.

Having considered ACRS' point of view, the AEC proceeded with its own agenda, installing a less robust confinement rather than containment and applying other standards outside commercial norms. From 1958 to 1974 the ACRS repeated its calls for improvements; it reiterated that the "confinement philosophy would not work," suggesting that reliance on confinement would result in consequences far greater than for contained plants in the event of a major accident. Such admonitions failed to overcome the pressing "national interest" focus on weapons production and the belief that accidents were unlikely.

Savannah River replicated ACRS' Hanford experience. The Atomic Energy Commission did not engineer the South Carolina plants to protect against serious internal accidents. Rather, concern focused on minor internal blasts and releases against which operators improved seals over the years. Plant design allowed for emission of some radioactive material, such as inert gases, during serious accidents.

In its role as institutionalized bureaucratic critic, ACRS raised numerous concerns: In the early years the danger of uncontrollable bursts of power, adequacy of confinements, and problems of power generation beyond design criteria topped the list. Later ACRS added emergency core cooling and seismic integrity. By 1960 it called upon the AEC to implement a better confinement as a matter of "urgency." To stimulate action, ACRS increasingly pointed to the burgeoning and increasingly safety-conscious commercial nuclear industry as an example. In a 1965 letter to the chairman of the AEC, the ACRS wrote, "... studies should be undertaken at an accelerated pace to determine means by which the above objectives [application of commercial licensing criteria] should be accomplished."[6]

In the end AEC introduced palliatives. For example, at Savannah River through the 1960s and 1970s safety improvements included the addition of filters to the plant's ventilation system to trap airborne radioactivity, backup motors for exhaust fans, the sealing of reactor rooms, emergency diesel generators, and improvement in the emergency core cooling and reactor shutdown systems.

These measures were important, but such remedies only marginally got to the heart of ACRS objectives, namely comparability to commercial plants in operation and design, notably a heavy reinforced containment.[7] Shrouded by the flag of military necessity, AEC refused to budge.[8] It would take more than hypothetical calculations to promote ACRS doubts to the level of pressing issues. It would take an accident—but as fate would have it, not just any accident.

Unbeknownst to the public, ACRS, and even to decision makers in the AEC, the academic discussions about the probability, nature, and consequences of an accident were being played out in ignorance of what was happening in the facilities themselves. On September 30, 1988, congressional testimony and intensive press coverage by the *New York Times* revealed 30 years of mishaps at Savannah River.[9] The most serious ranged from a power surge in 1960 that nearly went out of control, to the melting of fuel rods in 1970 and 1971.[10] Notwithstanding the portent, decision makers in Washington were kept in the dark by local managers.[11] Former officials of the Atomic Energy Agency speculate that the failure to inform policy makers and the public may have been motivated by the desire, on national security grounds, to prevent the Soviets from knowing about U.S. difficulties in producing nuclear materials and/or to avert public panic over commercial nuclear plants.[12] But these accidents did not harm the general public, and it would take something more dramatic to force policy makers to reconsider operating procedure. The opportunity emerged with Three Mile Island (TMI) and later Chernobyl.

THE LESSONS OF TMI AND CHERNOBYL FOR THE NUCLEAR PRODUCTION FACILITIES

The civil nuclear industry that dominates nuclear power generation today traces its origins to the development of the pressurized water reactor in the U.S. nuclear submarine program. It is not without irony that ACRS and later the General Accounting Office (GAO) would point to these more "sophisticated" licensed power plants as paradigms, inasmuch as the first major accident to endanger the public took place at Three Mile Island, a relatively newly minted commercial facility. But for TMI, and more so Chernobyl, little would have been done to elevate safety at the production reactors. Thus, their story has relevance to what was to take place at Hanford and Savannah River.

As in the case of the weapons production reactors, safety was not the "pressing issue." The 1954 Atomic Energy Act paints the matter with a rather broad brush. This is understandable; at the time the United States had no civil nuclear power industry to speak of. Nuclear promoters such as Edward Teller acknowledged risks but dismissed them. As in the case of the production reactors, getting on with the task of nuclear generation was the pressing issue.

Still, government was not oblivious. Technological policy critics in the bureaucracy—government engineers and scientists—evaluated nuclear hazards in the late 1950s and early 1960s and raised troubling questions about the possibility that major accidents could breach heavy steel-reinforced concrete containments. Their doubts were not sufficient to force major changes in public policy.[13]

What the technobureaucrats lacked was political connections and a black and white case to support their views. Bringing nuclear safety onto the political agenda was difficult. Calculations of deaths and contamination on paper were hypothetical, not demonstrable. The industry's reluctance to bear the cost of remedies complicated matters.

Then there was the fog of conflicting expertise. Although safety questions found an airing in scientific publications and the press, senior officials in the Atomic Energy Commission concluded that a catastrophic nuclear failure was noncredible. The remarks in 1974 of Herbert Kouts, then assistant director of reactor safety at the AEC, were typical. Relying on the draft *Reactor Safety Study* (RSS), government-sponsored effort to establish the probability of major accidents, Kouts asserted, "... [T]here will never be a major accident in a nuclear power plant. The odds on a major catastrophe were ... 'one in one billion to one in ten billion years' for a given reactor."[14]

Peer review of the RSS later revealed important technical flaws. Policy entrepreneurs such as Frank von Hippel of Princeton University called for a reevaluation of risk. Although his intervention stimulated a revision of the RSS, government monitors at the newly created Nuclear Regulatory Commission remained complacent even as reports from the field raised troubling questions about equipment breakdowns and lapses of quality control. These anomalies were not sufficiently "pressing" to encourage radical rethinking of nuclear safety. The conviction that the possibility of severe nuclear accidents was noncredible dominated. It would take a serious accident to force open a "window of opportunity" for reevaluation.

In light of the American experience, it was not surprising that the Soviets, who sought to emulate the United States but with fewer technological resources, would devote no more attention to nuclear safety. With the same overconfidence in the technology, Soviet energy planners carried on with the same bravado as American nuclear industrialists. Public relations in both societies were remarkably similar. In 1983, *Novosti* reported that "present-day Soviet nuclear stations were the safest in the world." The probability "of nuclear radiation involving a radioactive discharge was less than one in a million; workers had as much chance of being struck by lightning."[15] Shades of Herbert Kouts!

The Soviet program was not without dissenters; their effectiveness is another matter. It is not clear how items reach the Soviet political agenda. Given the difference in political cultures, analogies to the United States must be treated cautiously. Still, there are interesting parallels. Soviet scientists, like their American counterparts, air their concerns in professional publications and the press. Writing in 1979 in the important theoretical and political journal *Kommunist*, two academicians questioned the ecological costs of nuclear energy.[16] They suggested that such plants be concentrated in very remote regions outside European Russia. In 1982, a nuclear engineer working at Balakovo reported in *Sovetskaia Rossiia* that the plant received cooling pipes that were "complete junk."[17] Such comments about quality control appeared increasingly in Soviet literature and were analogous to critiques of both the civil and the production reactors in the United States.

It is uncertain to what extent such statements influenced decisions. They suggest that people in the field concerned about safety tried to get the ear of the authorities. But, as in the United States, events, not complaints, altered policy. In 1976, A.M. Petrosiants, head of the Soviet State Committee on Atomic Energy, informed an American counterpart that the heavy, reinforced concrete containments that envelop American pressurized water reactors (PWR) were unnecessary. He could easily have been speaking for the U.S. Department of Energy.

Reconsideration occurred after Three Mile Island, although the Soviets still exhibited no sense of urgency. They did not retrofit operating reactors with containments, although some new PWR construction included that protection. Cost, bureaucratic inertia, the low probability of accidents, and the negligible environmental consequences of TMI probably contributed to the decision. It took an accident of much greater magnitude to reverse such thinking and make nuclear safety a pressing issue. The following remarks from the Soviet report to the International Atomic Energy Agency in 1986 capture the sentiment that the Chernobyl accident overcame:

> The developers of the reactor installation did not envisage the creation of protective safety systems capable of preventing an accident in the presence of premeditated diversions of technical protection facilities and violations of operating regulations which occurred, since they considered such a set of events impossible.
>
> An extremely improbable combination of procedure violations and operating conditions tolerated by personnel of the power unit thus was the original cause of the accident.[18]

Finally it dawned on Soviet decision makers that the noncredible was in fact credible.

THE IMPLICATIONS OF TMI AND CHERNOBYL FOR U.S. PRODUCTION REACTORS

For the civil nuclear industry in the United States—which was already in trouble, having received not a single new reactor order since 1975—Three Mile Island was a commercial disaster from which it never recovered. As a result, no electrical utility commissioned new construction;

many plants under construction were cancelled. By contrast, the public paid little attention to military production reactors. DOE's response resembled that of the Soviets after TMI: it can't happen to us because we use a different reactor design.

I do not wish to suggest that TMI had no impact on DOE's weapons production facilities. But because public attention focused on commercial reactors, and because ACRS no longer bore responsibility for oversight, having shifted its mission to the Nuclear Regulatory Commission following the breakup of the AEC, DOE had no institutional critic to hold its feet to the fire. Thus, when DOE released an evaluation of the production reactors by the ad hoc "Training Committee" (known as the Crawford Committee report) in March 1981, it had little clout.

The report's conclusion is a curious mixture of sharp criticism and timidity. In its own words, "While the Committee found no evidence that the reactors operated in an unsafe manner, the number and type of deficiencies noted is cause for a number of concerns."[19] Deficiencies included the following.

- "Many of the 'TMI Lessons Learned' have not been adequately addressed or applied in DOE reactor programs."[20]
- "[O]verall there has been a significant diminution of the technical management capabilities of DOE organizations responsible for carrying out nuclear reactor programs when compared with their AEC organizational analogues . . ."[21]
- "At this point no safety requirement appeared to be mandatory for DOE reactor contractors."[22]
- "[T]he standards-making efforts at DOE have undergone substantial diminution both in emphasis and level of effort. DOE has opted to use NRC requirements as guidance. While using such requirements has merit, the manner by which DOE has done this has tended to cause confusion among DOE field offices. Furthermore effective DOE liaison has not been established with NRC standards organizations and relevant NRC documents are not routinely distributed within most DOE reactor program organizations. Many of those which are provided are not issued as requirements but guides."[23]
- "DOE lags behind the commercial industry in issuing uniform, unambiguous requirements for the selection, training and qualifications for reactor operating personnel. TMI [demonstrated] the importance of operator knowledge about heat transfer, fluid mechanism

and thermodynamics. The Committee considers DOE's effort to upgrade is inadequate."[24]

- "DOE's Secretary paid little attention to nuclear reactor safety prior to TMI."[25]
- "Technical capability in the nuclear program division and in the field offices varies widely and generally is inadequate and in some cases weak."[26]

Thus, notwithstanding its conclusion that DOE had not learned the lessons of TMI and had allowed safety standards to atrophy, the Crawford Committee gave the production reactors a passing grade: the plants were not "unsafe." Obviously, problems existed. Proposed remedies included improving competence in upper level management and technical capability at the field level, and setting safety guidelines, rather than requirements, for contractors to follow.[27]

In the years that followed, decision makers squandered the window of opportunity to make a significant departure in safety practices opened by TMI. DOE defensively took the position that the facts spoke for themselves: DOE reactors had an "excellent" record that required only marginal improvement.[28] Accordingly, by 1983, GAO reported that DOE had instituted little more than marginal improvements.[29]

Without institutional oversight, Congress tried to fill the post-TMI void through intensive use of GAO's auditing and investigative power to press for new safety measures. By virtue of this situation, in 1981, GAO assumed the role of an external quasi-institutionalized bureaucratic critic—an unofficial ACRS, so to speak.[30] In report after report, year after year, it criticized DOE for failing to give higher priority to safety institutionally, for not appointing oversight, and for not employing adequate expertise in the field.[31] GAO believed its efforts modified DOE behavior over time, despite a backlog of 198 inspectors' recommendations at Savannah River, some dating back to the 1970s, that had not been acted upon,[32] and 90 "Unusual Occurrence Reports" at Hanford, some of them serious.[33]

Seen from DOE, GAO's impact was minor. Speaking in frustration about GAO's ritual of presenting reports to Congress and its failure to influence policy, one internal bureaucratic critic, Paul W. Kaspar, DOE's acting Deputy Assistant Secretary for Operations, Defense Programs, testified in February 1988:

> I'm a little mystified as to exactly why we go through this year after year after year... We have 21 GAO reports here that point out different deficiencies

> in the system. They are all valid. They have not been refuted by anybody in DOE, and yet DOE has been slow to respond . . . Why were these GAO reports ignored for so long? We have got a lot of words and no action. You go out to these plants and look at them. Some of them are atrocious. They would not even begin to pass an NRC inspection. Why have we not done more? Is it expensive? I felt like Paul Revere up here because I have been telling people at every hearing that we have not fully addressed this problem yet. We are going to need tens upon tens of billions of dollars to clean up the mess created by 35 years of neglect.[34]

As GAO and internal bureaucratic critics tilled political vineyards, policy entrepreneurs attacked DOE in the courts. In one successful effort, nine environmental groups and three individuals living near Savannah River joined the state of South Carolina to sue DOE in November 1982 for failing to file an environmental impact report in the recommissioning of the L reactor, which had been out of service since 1968.[35] In July 1983 the effort bore fruit. The U.S District Court ordered the environmental impact statement. As a result DOE constructed a cooling pond. The department refused to implement additional safety improvements, however, and the plant resumed operation in 1985.

By 1985 the political and legal efforts to influence DOE clearly had become arduous. For the energy department, the pressing issue was not safety; it was, as it always had been, production of weapons grade material. When the L Reactor resumed operations, a substantial change in DOE behavior appeared remote.

Chernobyl turned the tables on DOE with a vengeance. Until the Soviet accident, DOE could dismiss incidents in the commercial sector that drew public attention as marginally relevant, since the department relied on a different design. But Hanford's N Reactor bore uncommon similarities to Chernobyl in terms of size and engineering, including a large graphite moderator situated in a building protected by a confinement. Furthermore, some analysts believed—incorrectly, as it turned out—that Chernobyl or its sister plants produced materials for the Soviet weapons program. Now commercial reactor operators could sit on the sidelines as the spotlight turned to Hanford and Savannah River.

DOE mounted a damage control effort, relying on the standard argument that the pressing issue remained national security, not safety, but the matter of safety of course could not be ignored. Although acknowledging in congressional testimony that Hanford shared similarities with Chernobyl, DOE accentuated differences, notably the N Reactor's negative void coefficient, lower operating temperatures, distinct fuel composition, and fast shutdown—scram—capabilities.[36] Nonetheless, the department indicated that it was prepared to introduce necessary remedies to assure safe operation of the plant.[37]

DOE's argument failed to win converts; rather it attracted the spotlight of the press, public interest groups, and Congress. In an effort to head off an early resumption of plant operations—Hanford had suspended generation in January 1987 for repairs—2 U.S. senators, 4 representatives, and 12 environmental groups, using the strategy employed in South Carolina in 1983, sought to enjoin DOE until it filed an environment impact report.[38] Perhaps more compelling were evaluations by a six-member panel chaired by Louis H. Roddis, Jr., and another by the National Academy of Sciences commissioned by DOE. The Roddis report included separate assessments by each consultant. Two, including that of the chairman, recommended closure of the N Reactor unless "requirements for defense material warrant accepting public hazards exceeding those of commercial reactors."[39]

The National Academy of Sciences released its appraisal in 1987. Refusing to rule on the plants' safety, "because such an analysis would invoke a determination of acceptable risk,"[40] the academy's findings clearly raised doubts that recalled years of criticism by ACRS, GAO, Congress, and environmental entrepreneurs. They included charges that

- DOE lacks a clearly defined safety standard.
- The reactors may be unable to compensate for a major loss of coolant.
- Confinement systems may not be adequate to contain large accidents.
- Safety oversight is ingrown and largely outside public scrutiny; DOE requires permanent safety oversight.
- DOE does not vigorously ensure that contractors implement safety orders.
- DOE fails to adequately audit contractor performance.
- DOE has failed to incorporate many lessons from TMI.[41]

What years of criticism failed to do, Chernobyl achieved. Finally, safety became the pressing issue. In 1987, DOE established in independent oversight advisory committee on nuclear safety chaired by John F. Ahearne, a former chairman of the Nuclear Regulatory Commission. In February 1988, after evaluating the cost of maintenance and operation and the declining national need for plutonium,[42] DOE placed the N Reactor in "cold shutdown." It preceded the decision by executing $100 million worth of repairs, and reserved the right to start up the plant in the future should a national emergency require doing so.[43]

To address National Academy of Sciences concerns over the adequacy of emergency cooling at Savannah River, operators, who had already reduced power by 26 percent in November and December 1986, cut power to 50 percent. In April 1988, DOE suspended operations to strengthen the plants' capacity to withstand earthquakes.[44] Finally, in August 1988, DOE proposed replacement of the Savannah River reactor with a new graphite-moderated plant and a smaller, high-temperature, gas-cooled reactor, a potential prototype for future commercial reactors, that was to be built at the Idaho National Engineering Laboratory.[45]

If the story ended with the announcement that the United States intended to replace the old, unsafe reactors, we might be consoled. We could tell ourselves that the federal government made economically costly mistakes—the Department of Energy calculates that the costs of plant modernization, environmental restoration and cleanup, and waste disposal might total $100 billion to $130 billion at the weapons production reactors and other plants that contribute to the manufacture of nuclear weapons[46]—but that the costs were only economic, since no accident that endangered the public had occurred. Unfortunately, events at Savannah River in August 1988 raised the specter of continuing danger. In what one DOE official called a "complete collapse" of safety procedure—and plant operators defined as a minor incident—an unexpected power surge exacerbated by operator negligence took place in the start-up of the P Reactor, which had just undergone four months of repair.[47]

DOE acknowledges that the episode illustrates that procedural sloppiness continues to infect operations.[48] The event is all the more disturbing because we may have to rely on the decaying[49] Savannah River plants for weapons material over the next decade. Unless safety remains the pressing issue on DOE's agenda, the roll of the dice may be what determines whether we shall yet witness a major accident.

THE TERRORIST THREAT

As we look to the future, there is another matter that ought to concern decision makers, namely sabotage and diversion of sensitive nuclear materials. Scanty information makes thorough assessment difficult, but what is available suggests that DOE must strengthen security.

The issue briefly emerged on the public agenda in congressional hearings in the mid 1980s, in part as the result of the efforts by a disgruntled energy department security employee, a whistle blower, John Hnatio.[50] In June 1983 Hnatio related to a member of the Senate staff his concerns about the adequacy of security at Savannah River and other facilities. He contended that the Department of Energy misled the White House on the matter in an annual report.

Resulting congressional investigations and hearings revealed that security was "a shambles." On the basis of mock attacks conducted by DOE's Independent Assessment Program and information from other sources, a House committee found

> examples of vaults containing significant quantities of plutonium in a desirable form which could not be defended, sensors and alarms which do not work, guards who cannot shoot. During one exercise guards fired on one another, rather than the attacking force. A guard force with less than one percent chance of interrupting an attacking force; a guard force that responded to a mock attack 16 minutes after the attackers had left with the plutonium; a Japanese camera crew, in a helicopter, filming a highly sensitive bomb facility without challenge; contractors refusing to allow guards to respond to mock attacks . . . a disgruntled former employee who drove past guards at the main gate and threatened to destroy a key facility . . .[51]

The evident sloppiness in security calls for an explanation. Hnatio found it in the presumption among DOE officials that the threat could not be defined and as a result did not exist. In Hnatio's own words, "I think indifference was the result of a widespread attitude that nothing was ever going to happen—a sort of 'it can't happen here' approach to security."[52]

Whether nuclear production reactors and adjacent special nuclear storage sites have been or will be attractive terrorist targets is open to debate. The Department of Energy, as distinct from the Nuclear Regulatory Commission, does not publish lists of incidents. Nonetheless, students of terrorism have suggested scenarios where terrorists, some of them state-sponsored, or disgruntled workers might attempt to sabotage facilities or divert sensitive material.[53] Events in the civil nuclear sector support this concern.

Between 1970 and 1984, out of a worldwide total of over 4,000 reported incidents of terrorism, 292 were acts or threats of sabotage or diversion of nuclear materials.[54] These included minor bombings (e.g., pipe bombs), diversion, sabotage, arson, threats of the same, and standoff attacks on reactors (e.g., a bazooka attack).[55] Perpetrated by terrorists, criminals, and hostile employees, the acts involved nuclear power plants, transport, and personnel. In recent years the frequency of such incidents has declined.

The question remains whether such incidents are likely to escalate in the future. History suggests that those having the greatest capability are the least likely to use the nuclear option, whereas the "mad bombers" are the least capable yet most likely to do so. After all, some groups have enough technical knowledge to kill large numbers of people by poisoning water supplies, sabotaging aircraft, or releasing gas, yet they do not. Their reluctance may lie in their assessment that such measures would be counterproductive, causing those they wish to influence to be repelled.[56]

However, individuals have engaged in acts of violence that most people would consider irrational. Evidence suggests that even "rational" terrorists have become bolder. The apparent sabotage of airliners in the mid 1970s by anti-Castro emigrés and Palestinians, resulting in more than 70 deaths in each incident, and the loss of hundreds of lives in the bombings of Air India and Pan American Airways flights in 1985 and 1988, respectively; the death of Iranians in a fire deliberately set in Iran in 1978; and the hundreds of casualties in the bombings of the U.S. Embassy and Marine compound in Beirut in 1983 point in this direction. Yet such incidents represent a tiny fraction of terrorist actions and it may be inappropriate to assume that they indicate a trend or that they point to the likelihood of nuclear terrorism. Still, given the uncertainties of human behavior, a terrorist act at a military or civil reactor cannot be ruled out. In the words of a recent RAND report, the absence of serious nuclear incidents to date should not "lead policymakers into a false sense of security."[57]

CONCLUSIONS

For those who seek responsible public policy, the story of nuclear safety and security at the Hanford and Savannah River nuclear weapons material production reactors is discouraging. The record shows that policy makers wedded to beliefs about safety that have been obsolete for decades were unwilling to shift gears, even in the face of mounting evidence that preserving the status quo risked serious harm to the public. The facts further suggest that the political process failed. Nine different administrations have occupied the White House since the production reactors went into operation. Not one made nuclear safety a priority.

Marginally effective oversight compounded problems. Although the

Advisory Committee on Reactor Safeguards, until 1974, and the General Accounting Office thereafter performed admirably in bringing tough questions to the attention of decision makers, their success was marginal. At worst, policy makers ignored them; they had no political clout. Although they had access to decision makers, they could not enforce recommendations. Political entrepreneurs, including environmentalists and other citizen groups, may have had limited access, but at times they seemed to have more influence through their use of the courts.

In the end, events, not politics, changed operations at the production plants. What is most surprising is the gravity of events that were required to move decision makers. To overcome the "accidents can't happen here" sentiment, the mishaps had to be catastrophic. Three Mile Island was insufficient. Had it not been for Chernobyl, the opportunity for reconsideration of America's nuclear weapons production complex would not have occurred. That it would take an incident of this magnitude to force U.S. reconsideration of nuclear safety supports the proposition—the "body in the morgue" approach—advanced years ago by John Gofman and Arthur Tamplin.[58]

These Cassandras—ultimately ineffective policy entrepreneurs—contended that the U.S. Atomic Energy Agency would only take remedial measures to make nuclear energy safe after people were hurt. In the aftermath of Chernobyl, the contention appears to have international validity. Technological overconfidence, bureaucratic conformity and lethargy, cost, pride, a preoccupation with secrecy, the drive to expand and maintain weapons production, military necessity, deceit by plant operators and local energy officials, a desire not to hurt commercial nuclear power, ineffectual political entrepreneurs and bureaucratic critics collectively explain why it took so long to make safety a matter of national priority.[59]

NOTES

1. For a discussion of the theory of public policy making see John W. Kingdon, *Agenda, Alternatives and Public Policies*, Boston: Little Brown, 1984.
2. National Research Council, *National Academy of Sciences, Safety Issues at the Defense Production Reactors: A Report to the U.S. Department of Energy* (National Academy Press, Washington D.C., 1987), pp. 1-2; Thomas Cochran, et al., *U.S. Nuclear Production* (Cambridge, Mass., Ballinger, 1967), pp. 52, 65-67.

3. The P, L, and K reactors originally went into operation in February, July, and November 1954, respectively. For information on their design characteristics see Gordon R. Thompson, Testimony, "Implications of Severe Reactor Accidents at the Savannah River Plant," Richland, S.C.: The County Council, Richland, South Carolina, pp. 9-12, available from Institute for Resource and Security Studies, Cambridge, Mass.
4. National Research Council, op.cit., pp. 2-4.
5. For a review of ACRS' failed effort to materially modify safety procedure at the production reactors, see memo from Victor Stello, Jr., to Nuclear Regulatory Commissioners Palladino, Roberts, Asselstine, Bernthal, and Zech, "Review of Files on 'N' Reactor and Savannah River Plant Reactors," July 30, 1986. Nuclear Regulatory Commission, Mimeo.
6. Ibid., p. 21.
7. For a safety profile of the Savannah River plants see Thompson, op. cit., pp. 8-14.
8. Stello, op. cit., pp. 20-23.
9. *New York Times*, Oct. 1, 1988.
10. See letter from Dupont scientist G. C. Ridgeley to G. F. Merz "SRP Reactor Incidents of Greatest Significance," August 14, 1985, available from the Senate Government Affairs Committee. See also Senator John Glenn, News Release, Sept. 30, 1988.
11. *New York Times*, Oct. 3 and 4, 1988. A 1984 Environmental Impact Report produced by DOE claimed there had never been fuel melting at Savannah River. See Energy Research Foundation, et al., Civil Action 88-3541 Complaint, U.S. District Court, District of Columbia, p. 9.
12. *New York Times*, Oct. 5, 1988.
13. For a discussion of these individuals see Daniel ford, *The Cult of the Atom* (New York: Simon & Schuster, 1982).
14. Ibid., p. 157. Note that this overstated the conclusions of the Reactor Safety Study, which put the risk of a significant accident at 1 in 20,000 reactor years.
15. Quoted in David Marples, *Chernobyl and Nuclear Power in the USSR* (New York: St. Martin's, 1986), p. 96.
16. Ibid., pp. 100-101.

17. Ibid., p. 107.
18. State Committee for Using Atomic Energy of the USSR, *Accident of the Chernobyl AES and Its Consequences,* Washington D.C.: U.S. Department of Energy, Aug. 17, 1986, p. 34.
19. Department of Energy, "A Report on A Safety Assessment of the Department of Energy Nuclear Reactors," DOE/US-0005, Washington D.C.: U.S. Department of Energy, np. Mar. 1981.
20. Ibid., p. 1.
21. Ibid., p. 15.
22. Ibid., p. 18.
23. Ibid.
24. Ibid., p. 21.
25. Ibid., p. 38.
26. Ibid.
27. Ibid., pp. 47-53.
28. Comptroller General, "Better Oversight Needed for Safety and Health Activities at DOE's Nuclear Facilities," EMD-81-108S (Washington D.C.; U.S. General Accounting Office, Apr. 14, 1982), p. 1, 3.
29. U.S. General Accounting Office, "DOE's Safety and Health Oversight Program at Nuclear Facilities Could Be Strengthened," GAO/RCED-84-50 (Washington D.C.: U.S. General Accounting Office, 1983).
30. The first major report following TMI was Comptroller General, op. cit. For a review of other reports, see "Reactors at the Savannah River Plant," Hearing Before the Subcommittee on Strategic Forces and Nuclear Deterrence of the Committee on Armed Services, U.S. Senate, 100th Cong. 2d sess., Feb. 29, 1988 (Washington D.C.: Government Printing Office, 1988).
31. Ibid.
32. Eliot Marshall, "How Safe Savannah River Reactors," *Science,* March 27, 1987, pp. 1563–64.
33. Letter from Daniel Hirsch, Director, Program on Nuclear Policy, University of California, Santa Cruz, to Representative Edward Markey, Aug. 28, 1986. Among the incidents in the "Unusual Occurrence Reports" were numerous malfunctions of the reactor shutdown mechanisms (e.g., major components of the control rod

and ball drop systems that fail to operate in the time required or fail to operate at all); damage to emergency core cooling system pipe supports; numerous pump failures; and damage to key piping and valves, which reduced the integrity of the emergency core cooling system. Letter available from Daniel Hirsch, Director, Committee to Bridge the Gap, Los Angeles.

34. "Reactors at the Savannah River Plant," Hearing Before the Subcommittee on Strategic Forces and Nuclear Deterrence of the Committee on Armed Services, U.S. Senate, 100th Cong, 2d Sess. (Washington D.C.: U.S. Government Printing Office, Feb. 29, 1988), pp. 21-22.
35. Daniel Charles, "The People v. the Complex," Dan W. Reicher and S. Jacob Scherr, "Laying Waste to the Environment," *Bulletin of the Atomic Scientists,* Jan./Feb. 1988, pp. 29-34.
36. Subcommittee on General Oversight, Northwest Power and Forest Management, Committee on Interior and Insular Affairs, U.S. House of Representatives," Statement of M. J. Lawrence, manager, Richland Operations Office, U.S. Department of Energy," May 19, 1986, mimeo. For another assessment, see U.S. General Accounting Office, "Nuclear Safety: Comparison of DOE's Hanford N-Reactor With the Chernobyl Reactor," GAO/RCED-86-213BR (Washington D.C.: U.S. General Accounting Office, Aug. 1986); *Congressional Record*—House, p. H 4798, July 23, 1986.
37. Letter from Mary L. Walker, assistant secretary, Environment, Safety and Health, Department of Energy, to the undersecretary, Dec. 10, 1986.
38. *New York Times,* July 18, 1987.
39. Letter from Louis H. Roddis, Jr., to John L. Herrington, secretary, U.S. Department of Energy, Oct. 3, 1986.
40. National Academy of Sciences, op. cit., p. xiii.
41. Ibid., see entire report.
42. The staff of the Northwest Environmental Advocates brought these rationales to my attention in a telephone interview, Oct. 4, 1988.
43. *New York Times,* Feb. 17, 1988. The most notable repairs were made to the plant's seismic integrity and ability to handle hydrogen explosions. Two years of preparation would be required for the plant to resume operations.

44. Even with the upgrading, the plants will not have the ability to withstand earthquakes that is comparable to commercial reactors, Energy Research Foundation, op. cit., p. 11.

45. Irwin Goodwin, "DOE Wants New Weapons Reactors to Replace Aging, Troubled Ones," *Physics Today*, Sept. 1988, pp. 47-48; J. Raloff, "Reactor Troubles Highlight Safety Concerns," *Science News*, Aug. 27, 1988; *Washington Post*, Aug. 19, 1988. Another option under consideration in late 1988 and 1989 was construction of a $2.3 billion linear accelerator to produce tritium. *New York Times*, Jan. 23, 1989, pp. A1, A9.

46. *New York Times*, Sept. 22, 1988; Department of Energy, Environment, *Safety and Health Report for the Department of Energy Defense Complex* (Washington D.C.: U.S. Department of Energy, July 1, 1988). The National Academy of Science estimated that modification of the Savannah River plants would cost over $500 million. Energy Research Foundation, loc. cit. Senator John Glenn puts the estimate for the total cleanup at $150 to $175 billion. See Senator John Glenn, News Release, Sept. 30, 1988.

47. Irwin Goodwin, "DOE Wants New Weapons Reactors to Replace Aging, Troubled Ones," *Physics Today*, Sept. 1988; J. Raloff, "Reactor Troubles Highlight Safety Concerns," *Science News*, Aug. 27, 1988; *Washington Post*, Aug. 19, 1988; U.S. General Accounting Office, "Ineffective Management and Oversight of DOE's P-Reactor at Savannah River, S.C., Raises Safety Concern," GAO/T-RCED-88-68, Sept. 30, 1988; Forrest J. Remick, "No 'Runaway,' No Cover-up at Savannah River," [op-ed], *Wall Street Journal*, Dec. 13, 1988, p. A 20; letters from D. A. Powers to E. Schwartz, Sandia National Laboratories, Aug. 24, 1988, R. M. Keller, to R.W. Starostecki (DOE Deputy Assistant Secretary Safety, Health and Quality Assurance), n.d., available from the Senate Government Affairs Committee. For more recent evidence of managerial sloppiness see, *New York Times*, Jan. 25, 1989, p. A8.

48. Richard W. Starostecki (DOE deputy assistant secretary Safety, Health and Quality Assurance), "Memorandum: Savannah River (SR) PO Reactor Events—Safety Implications," U.S. Department of Energy, Sept. 16, 1988. In this memo Starostecki states, "There are currently some senior managers within the Department with an attitude toward production reactor safety, which on the face seems to

be similar to that which existed in the space program prior to the Challenger accident." p. 2.

49. The decay focuses on cracks found in critical cooling pipes. *New York Times,* Dec. 21, 1988, pp. A1, A13.

50. "Nuclear Security Cover-up," Hearing Before the Subcommittee on Oversight and Investigations of the Committee on Energy and Commerce, U.S. House of Representatives, 98th Cong., 2d Sess. (Washington D.C.: Government Printing Office, Feb. 3, 1984).

51. Ibid., p. 2.

52. Ibid., p. 51.

53. Paul Leventhal and Yonah Alexander, eds., *Preventing Nuclear Terrorism,* Lexington, MA: Lexington Books, 1987; Paul Leventhal and Yonah Alexander, *Nuclear Terrorism: Defining the Threat,* Elmsford, NY: Pergamon-Brassey, 1986; Louis Rene Beres, *Terrorism and Global Security: The Nuclear Threat* (Boulder, CO: Westview Press, 1987).

54. Bruce Hoffman et al., "A Reassessment of Potential Adversaries to U.S. Nuclear Programs," R-3363-DOE (Santa Monica: Rand Corporation).

55. Gail Bass and Brian Jenkins, "A Review of Recent Trends in International Terrorism and Nuclear Incidents Abroad," N 1974-SL (Santa Monica, CA: Rand Corporation 1983), p. 26.

56. Brian Jenkins, "Will Terrorists Go Nuclear?" Discussion Paper No. 64 (Los Angeles: California Seminar on Arms Control and Foreign Policy, 1975).

57. Bruce Hoffman, op. cit. p. 7.

58. Ford, op. cit., p. 206.

59. For a viewpoint more sympathetic to DOE's management of Savannah River see Remick, loc. cit. For more information see Energy Research Foundation, "Reactor Safety at Savannah River Plant: A Chronology, 1978-January 1990," *Research Notes,* Columbia, SC: Energy Research Foundation, 1990.

5

On the Road to Ruin: The Transport of Military Cargoes

GERALD JACOB

ANDREW KIRBY

IN THIS CHAPTER, OUR AIM IS TO SHOW THAT THE CIRCULATION OF MILITARY cargoes around the United States poses very real dangers to the population and has a number of serious social, economic, and political implications. This country's strategic concerns have led to a situation in which facilities for the production of weapons are dispersed throughout the states, and there is a constant parade of materiel from one location to another. This circulation is open to the possibility of an accident, as is any movement of hazardous or radioactive materials taking place by road or rail. Many of

The authors would like to thank a number of persons. Andrew Knudtsen undertook a great deal of valuable research; Tom Wiselogel and Alison Habel drew the map; Kenneth Kirby kept us informed about developments in Europe. The excerpt from White Noise *is courtesy of Don DeLillo and Viking Penguin. Thanks are also due to Stephen Schneider for facilitating this research. This paper is a joint effort for which the authors alone are responsible: the views expressed here should not be attributed to any of the institutions with which they are, or have been associated.*

these military cargoes—which include munitions, ordnance, fuels, delivery systems, radioactive products and waste, chemicals, and other raw materials and components used in the assembly of conventional and nuclear weapons—are, by definition, highly dangerous. In short, there is a real risk of accident and environmental ruin.

These dangerous cargoes would be enough to constitute an important footnote in the annals of everyday risk. The story, however, is much more complex. In the first instance, shipments take place within a miasma of secrecy as these cargoes are moved from the closed spaces where military law is absolute, into the world where more complex political relations exist. Powerful steps are taken to wrap this transportation within the envelope of military privilege. Although states and localities face potential threats from these cargoes, they often have little involvement in overseeing shipments and are rarely informed of routing or scheduling. Even as peace campaigners and residents attempt to understand the movements of these cargoes, efforts continue to preserve secrecy and to exclude citizens from learning more about the military's business.

The difficulty of assembling data on the nature and movement of military cargoes and accidents arising therefrom makes it difficult to assess their current environmental consequences and the potential liabilities of this transportation. Furthermore, these cargoes are subject to a wide range of bureaucratic and regulatory regimes, monitoring, and reporting requirements, which make it even more difficult to assemble a comprehensive view of this traffic. The situation with respect to weaponry is thus different from nuclear power, where commentators have pointed to the creation of a "nuclear state"—a set of powerful institutions that control the production of nuclear power and undertake to improve the public's tolerance of the industry.[1] In the latter case, a good deal of effort has been put into reassuring the country that nuclear power is benign. As far as actual weapons are concerned, it would of course be hard to suggest that they are "safe."[2]

Because of secrecy and the difficulty of assembling reliable data on the movement of military cargoes, we must often rely on analogy when attempting to assess the potential consequences of this traffic. Fortunately, much can be inferred about the transport of military cargoes by examining the analogous movements of hazardous and radioactive materials within the civilian sector. And even though military and civilian cargoes may be subject to very different political regimes, many of the problems created by the transport of hazardous civilian cargoes can be associated with military cargoes as well. Throughout this chapter, then, we will as necessary examine the commercial transport of hazardous cargoes for the insights these analogous cases provide into the potential consequences of military transport.

RISKS AND DANGERS

At this point it is important to elaborate on interpretations of risk. Throughout much of social science, risk analysis is a formalized process that hinges upon the identification of probabilities. By this logic, nuclear power was, until Chernobyl at least, argued to be safe because there were no known deaths associated with its generation, and the risks of irradiation appeared low. Similarly, the transport of radioactive cargoes (e.g., spent reactor fuel) has been argued to involve low probabilities of excessive irradiation or an accidental release of radioactive material. (We will focus on commercial nuclear power for the moment, as a large amount of research is available on the subject.) The U.S. Nuclear Regulatory Commission (NRC) estimates that the transport of radioactive materials is innocuous[3] and that the probability of exposure to a dangerous dose of radiation from a transportation accident is negligible.

Instead of measuring the probability of an accident in terms of *past* conditions, a different approach to risk would proceed on the assumption that a serious accident involving a military cargo will occur some time in the *future*. This assumption is not unreasonable since typical annual data on commercial hazardous materials accidents show more than 15,000 incidents per year—at a cost of several hundred injuries, more than a dozen fatalities, and over $10 million in property damage. Official reports may actually capture a small proportion (perhaps only 10 percent) of the total number of incidents involving hazardous materials.[4] Thus, while we do not discount past experience, we are skeptical about relying on such experience and subsequent estimates of probability as a substitute for vigilance. This point is reinforced in that past experience may say relatively little about the safety of transporting newly invented weapons systems, packages, and components, or exotic new chemicals, biocides, and waste products.

As an alternative approach for assessing the magnitude of the risks, the potential extent of problems created by the transport of commercial nuclear waste can be calculated in terms of the population residing within transportation corridors: that is, the number of individuals exposed to shipments. This approach does not depend upon the usual calculation of risk probabilities; the nation's experience with nuclear waste transport is limited, and the latter is nowhere near the scale of shipments contemplated for the future. This alternative approach begins with the assertion that an accident may occur at any point along a transportation route. The assumption is supported by the Department of Energy's (DOE) intention to provide training to all emergency response personnel along all likely routes to a high-level nuclear waste repository.

How would individual households respond to such an accident and with what results? Studies of other situations involving nuclear materials show that safety margins are defined very differently by citizens and by professional emergency management personnel. Compared with official evacuation zones, self-evacuation may involve households in a much wider distance band around a nuclear accident, and may even continue up to 25 miles from an accident site.[5] Given that past evacuations involving nuclear materials have also included locations up to 25 miles from an accident, it seems safe to estimate that all residents of a county through which a designated truck route passes may potentially view themselves as being at risk from a transportation accident.[6] On this basis, approximately 80 million persons, or nearly one-third of the popluation of the United States could regard themselves as being at risk from road accidents involving nuclear materials. This is not an estimate of the population that would be affected by a single incident; rather, it represents a group that could be *concerned* about the transport of such cargoes. This estimate also suggests that a large proportion of the U.S. population must be prepared to deal with the consequences of a nuclear waste accident. Moreover, given that the Nevada nuclear waste repository (under the control of the DOE Office of Civilian Radioactive Waste Management) will also be used for military waste disposal, the potential exists for this same number of people to be concerned about the consequences of transporting military cargoes.

The grounds for such concern have become more clear with the growing inventory of past accidents. A prototypical incident occurred in Denver, on August 1, 1984, when a truck carrying naval torpedoes overturned at 4:50 a.m., while travelling through Denver's busiest interstate highway interchange. Emergency personnel had no warning. The truck was not even scheduled to be in Colorado. Although the vehicle was posted with 24-hour telephone numbers to be used in case of accident, there was no answer to repeated calls. Military personnel did not reach the accident site until 9:04 a.m. Eventually, it took eight hours to remove the potential dangers from Denver's highways.[7] This incident involved relatively unsophisticated weapons technology; even so it caused serious disruption and evacuation, and required a complex interagency emergency response. We must presume that incidents involving nuclear weapons, or a toxic product like Agent Orange, could cause equally serious disruptions.

From such cases, we make the following observations: First, a road or rail accident involving military cargoes could result in widespread self-evacuation, and produce social and economic disruption, financial costs for local jurisdictions, and emergency response problems. Second, as such threats become discussed more widely in the press and other forums, public concern and opposition of the "no transport through my backyard" variety will become commonplace. This has already occurred in towns and cities along transportation routes for hazardous materials[8] and it will occur regardless of risk assessments that show the probability of an accident to be low and the environmental consequences of an accident to be insignificant. Concerned citizens living within transportation corridor routes may demand both unlimited liability coverage and compensation from waste transporters, and will also attempt to influence the management of all shipments of hazardous materials. Finally, the public may make few distinctions between the transport of hazardous commercial or military cargoes, nuclear or chemical cargoes, military or civilian nuclear waste, and the levels of risk associated with different materials and forms of transport. In short, barring evidence to the contrary, skepticism about the safety of one type of transport may easily be transformed into skepticism about the safety of transporting military cargoes.

Methods of analysis can have a major influence upon the way in which a problem is understood. In terms of risk probabilities, the possibility of accident may appear insignificant. Seen through our alternative approach—which asks, what if?—the transportation issue has the potential to involve a significant proportion of the country's population. We may of course be wrong, and we hope that we *are* wrong; it is possible that there will never be a catastrophic traffic accident involving military cargoes. However, that does not diminish the gravity of the issue, the importance of public concerns, or the numbers of people who may demand participation in political processes governing military transport.

THE PERVASIVENESS OF THE TRANSPORT OF MILITARY CARGOES

A basic characteristic of the circulation of the raw materials and products of war-making is its *pervasiveness*. By this we mean that the large volume of weapons-related cargoes can be difficult to distinguish from the transport of everyday goods and materials. The demands placed by weapons-producing facilities upon political and economic institutions, and the subsequent consequences of these activities, are woven into the structure of everyday life. Moreover, the ubiquity of weaponry leads us to a number of other issues: accountability, the demands for prior knowledge of shipments, the extent of federal preemption, and the local costs of safety.

Two general geographic realities emphasize the pervasiveness of the traffic in military cargoes. First, the transportation system of this country joins urban centers. Interstate and other major highways pass near central business districts and through densely populated areas. For example, the route used to transport radioactive materials from Brookhaven National Laboratories (which is involved in weapons-related research) passes through New York City, where population densities approach 100,000 per square mile. Most railroad lines, for historical reasons, lead to terminals and yards adjacent to central business districts.

A case taken from a 1983 accident report of the National Transportation Safety Board, although not involving a specifically military cargo, exemplifies the vulnerability of urban centers to rail accidents involving hazardous materials. At 4:00 a.m. a railroad tank car carrying nitric acid ruptured and released a cloud of toxic gas within three miles of downtown Denver. Two interstate highways were closed, 9,000 residents were evacuated, and area businesses were forced to remain closed. Despite such accidents, the transport of hazardous materials through densely populated areas is common. Indeed, it is often seen as unavoidable, given that rail routing options are limited. In the case of highway transport, beltways and bypasses may be used to reduce exposure but they are not a panacea. Diurnal changes in the distribution of urban populations mean that cargoes travelling suburban beltways and bypasses at night may pose greater risks than if a central city route were used.

The pervasiveness of the traffic in military cargoes is reinforced by a second reality: that weapons industries have been dispersed throughout the United States. There are at least three reasons for this. Originally, nuclear facilities were dispersed geographically, in part to maintain secrecy and to foil sabotage; the Manhattan project of the 1940s is the quintessential example. Scientists were gathered in then remote locations such as Hanford, Washington, and Oak Ridge, Tennessee, each to work on a discrete part of the project, unknown to the others. Today, the legacy of those decisions remains, as the manufacture of nuclear weapons continues to be carried out at these locations. Secondly, many manufacturing companies were decentralized during World War II for strategic reasons. As a result, a large number of then small Sunbelt cities became the sites of military plants.[9] Finally, dispersal has been abetted by the political process, military production being a seemingly bottomless pork barrel that legislators can use to promote economic development within their districts. Conversely, a military facility or contract is the reward an agency can bestow on the district of a supportive congressional member. As a consequence, weapons plants are scattered throughout the country, their operations now unavoidably connected by a tangle of transportation routes. The map (figure 5.1) illustrates this network for nuclear weapons facilities only.[10] It would be hopelessly complicated to include all conventional weapons facilities as well.

Dispersal of military production means that raw materials and partly finished weapons components must be shuttled among numerous facilities. Several transcontinental shipments may be required for the completion of a product, with each stage of production linked to the next via a rail car or a flatbed trailer on an interstate highway. The production of a single nuclear weapon may involve shipments of radioactive materials to and from Hanford or Savannah River (where reactors produce plutonium); Rocky Flats, Colorado (where triggers for weapons are fabricated); the Pantex plant in Amarillo, Texas, where final assembly takes place; missile sites in Wyoming, North Dakota, or Europe; the Idaho National Engineering Laboratory, where old warheads and spent fuel are reprocessed; and, finally, a nuclear waste site in New Mexico or Nevada. A similarly extensive movement of materiel is involved in the manufacture of chemical, biological, and other weapons.

ACCOUNTABILITY

Concerns about poor oversight over the transportation of commercial hazardous and radioactive materials have increased in the last few years, but transportation of these same cargoes for military reasons has escaped

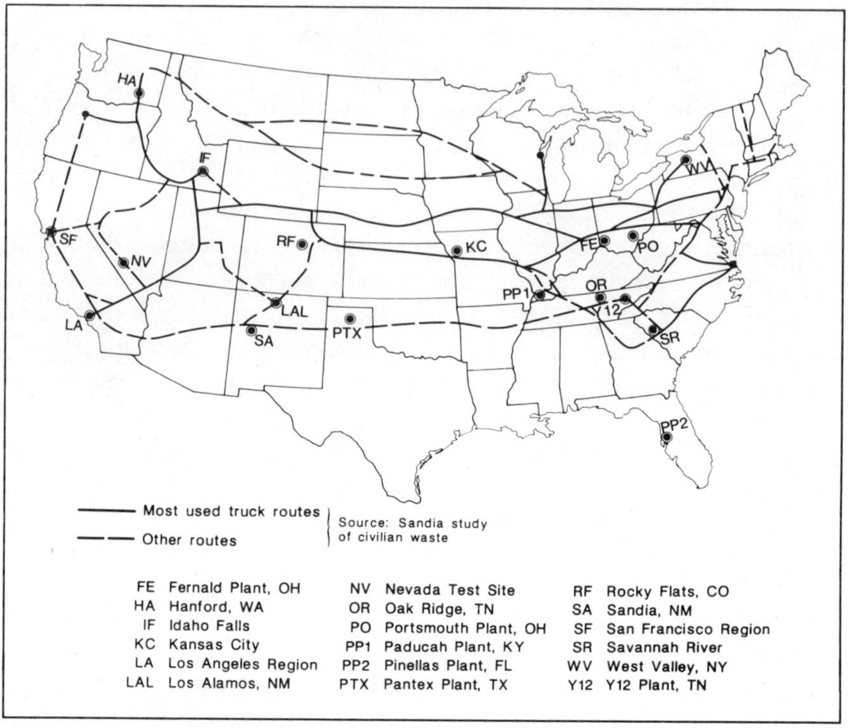

Figure 5-1: Waste Shipping Routes

that level of scrutiny. For example, commercially used radioactive materials must be transported in a package whose design is certified by the Nuclear Regulatory Commission (NRC). (It should be noted that the NRC certifies the design, but not each package that will be used.) As part of the design certification process, manufacturers submit evidence that the package will withstand a set of accident conditions. Casks used to transport spent fuel from commercial reactors, for example, must be able to withstand a 30-foot drop onto an unyielding surface, a puncture test, a 1475°F fire for 30 minutes, and an immersion test. The NRC reviews the evidence to assess whether a cask does indeed meet these standards, and the record of NRC deliberations is open for public review. NRC-licensed facilities, such as commercial reactors, must use certified packages for the transport of radioactive materials.

In its *civilian* nuclear program, the Department of Energy maintains that the cask is the major defense against a release of radionuclides in the event of a transportation accident. The agency points to NRC certification when assuring the public that spent reactor fuel can be transported in safety. Relying on its confidence in NRC-certified packages, DOE contends that the probability of a catastrophic release from the latter is so low that the materials can be transported safely by the railroads on general freight trains. In addition, DOE and the NRC contend that it is unnecessary for these shipments to be accompanied by guards, emergency response teams, or radiological monitors.

However, radioactive materials transported to and from DOE's weapons facilities—which are not subject to NRC licensing or oversight—need not be carried in NRC-certified packages. Congress has permitted DOE to "self-certify" these packages. In 1988 the General Accounting Office (GAO) reviewed the effectiveness of DOE self-regulation. In its report, *DOE Needs to Take Further Actions to Ensure the Safe Transportation of Nuclear Materials,* GAO found that

> DOE performs the certification and maintains the certification files at the Albuquerque Operations Office for the packages it uses to transport nuclear explosives, nuclear components, and special assemblies (collectively known as "nuclear weapons packages"). GAO found three package designs that were used for several years without ever being certified and four package designs that had only 60-day approvals but had been available for use from 9 to 11 years... GAO then judgmentally selected 14 of the 42 package designs for a more detailed review and found that 7 of the 14 files did not fully demonstrate that the packages met standards equivalent to those prescribed by the Nuclear Regulatory Agency, as required by DOE policy.[11]

Whatever one's opinion of the adequacy of NRC regulations and standards, the GAO findings were that DOE's certification process for nuclear weapons packages lacked even that level of assurance and credibility.

Numerous packages that the NRC has judged unacceptable for commercial transport have been used by DOE for military shipments. (It was a noteworthy exception when DOE agreed to seek NRC certification for the TRUPACT II, the package that will be used to transport transuranic waste generated by weapons programs to the Waste Isolation Pilot Plant [WIPP] in New Mexico.) After using some packages for military shipments, DOE has submitted them to the NRC with the request that they be approved for commercial shipments. In at least nine cases, the review process was stopped after NRC questioned the packages' design.

The history of the MH-1A cask is illustrative: originally fabricated in 1971, it was used to ship military reactor fuel. In 1982, hoping that NRC licensees could also use the container, DOE requested NRC approval. The commission raised numerous questions about the cask, however, including the adequacy of the fire shield. After some modifications to the cask, DOE used it in 1985 for 13 shipments from Brookhaven National Laboratory, near New York City. (Packages used for such shipments in DOE's "experimental programs" also need not be NRC certified.) Despite these alterations, the NRC was not convinced that the container met the standards governing commercial shipments, and again called on DOE to address these concerns. When DOE did not respond, the U.S. Department of Transportation ordered DOE to suspend use of the container. Later, DOE said it still intended to pursue licensing, but a year after the Brookhaven shipments, it evidently had difficulty demonstrating that the cask could survive the standard drop test. Thus, although this container was not denied NRC certification, active solicitation of such certification appeared to end after the NRC raised questions about the integrity of the container. The GAO has identified other cases of uncertified or inadequately tested nuclear weapons packages that were used for nuclear weapons assemblies and explosives. It also found that DOE continued to use packages for transport even after potential defects in those packages were identified.[12]

To add to the pessimism such cases create, the Department of Defense appears less responsive than private industry to demands that it upgrade the safety of transportation vehicles. Since 1983, inspectors of the Federal Railroad Administration have found hundreds of defects in the aging and poorly maintained freight and tank cars used to haul military cargoes.[13] The Military Traffic Management Command, which directs a fleet of 1,185 cars, was found to be years behind in maintenance and inspections. In spite of ballooning military budgets during the Reagan administration, the Command claimed that its fleet of tank cars had deteriorated because of budgetary constraints.

Notification of an impending shipment of a hazardous or radioactive material is an important way to enhance local communities' emergency preparedness. Prenotification permits states and local communities the options of escorting shipments, alerting emergency response personnel, checking a carrier's compliance with vehicle safety rules, and otherwise monitoring a shipment's progress through an area. For its licensees, the NRC requires that states be given prenotification of highway shipments involving specified quantities of radioactive materials. DOE facilities—with the exception of the planned high-level waste repository—are not legally obligated to provide such notification. However, responding to demands from states that it adopt prenotification rules for shipments of military radioactive waste, DOE recently adopted a policy whereby it agreed to use NRC standards for all except classified shipments of nuclear materials.

This policy means that states will now receive prenotification of hundreds of shipments of transuranic waste (which include plutonium) from military facilities to the WIPP plant in New Mexico, as well as shipments for DOE research purposes. Classified shipments of nonwaste products, however, remain uncovered by this policy. In the future, it is possible that satellite tracking of shipments may replace written prenotification, provided states can afford the costs of maintaining their link with such a system. This system would give the capacity for 24-hour monitoring of shipments and quick access to emergency response information. DOE's prenotification policy is, however, an internal one. The policy could be changed again in response to pressures from within DOE's military programs, or in response to concerns about citizen action against shipments. The latter could result in more shipments being designated as classified, in the "interest of national security," and thus not subject to prenotification.[14]

INTERSTATE COMMERCE AND PREEMPTION

Should the same arguments made for removing barriers to interstate commerce in peanut oil be extended to the traffic in hazardous and radioactive materials used for weapons production? Can the transport of nuclear weapons and military cargoes ever be considered safe? Should the public be assured by a DOE spokesman who, in a variation on a scene from *Dr. Strangelove*, said a cask of high-level radioactive waste from Three Mile Island was so safe to transport that "if I had a saddle, I'd ride that SOB across the country"? Bravado aside, most Americans know little about the hazards and quantity of specific cargoes being transported through their communities. Extensive research is required just to identify what is being moved. The Environmental Policy Institute found that in 1987, for example, 11 million residents were being endangered by the transport of a toxic rocket fuel component, nitrogen tetroxide, from Vicksburg, Mississippi, to Martin-Marietta laboratories in Denver and other locations, including Vandenberg Air Force Base in Los Angeles and Aerojet General Systems in Sacramento, California. This highly toxic product was being shipped irregularly, and without the knowledge of these communities, for years.[15]

Again, the pervasiveness of military production in the U.S. economy often makes it difficult to distinguish between cargoes destined for civilian and for military purposes. Shipments of materials for the U.S. weapons complex should be evaluated in light of the issues raised by the transport of commercial radioactive and hazardous materials. Thus, examples taken from commercial transport are invaluable for suggesting questions about the transport of cargoes serving the military and weapons complex.

The Interstate Commerce Commission has ruled—despite railroad companies' objections—that railroads could not require that highly radioactive materials be carried on special trains. According to the ICC, railroad cars carrying casks of spent reactor fuel and other highly radioactive materials could be safely hauled by regular freight trains like any other cargo—largely because of the types of containers used to transport radioactive materials. Special trains are more expensive because they are operated under special procedures, such as speed limits. The ICC found that the added expense of such trains created no parallel increase in safety or protection. Nevertheless, many classified military shipments of radioactive and other materials are carried on special trains or on trains accompanied by guards. It is obviously difficult to obtain the information necessary to assess whether shipments that the ICC (and the Department of Energy) claims should be moved by regular freight trains pose any less danger than the classified shipments moved by special trains.

Railroad companies are under no special obligation to inform local communities about hazardous materials being shipped through the area. Companies will provide information on the carloads per year of various commodities they carry, broken down according to the U.S. Department of Transportation's commodity classification scheme. In 1986, for example, the Burlington Northern Railroad carried 96,293 carloads of hazardous materials, including 672 carloads of explosives, 14,180 carloads of flammable gas, 179 carloads of radioactive materials, and 4,736 carloads of chlorine. However, railroads will seldom provide reports of commodity traffic along specific routes or through specific communities. The flow of hazardous materials carried by truck is also difficult to assess. States have had to conduct surveys at ports of entry just to estimate the types and quantities of hazardous cargoes being carried into their jurisdiction.

The carriage of a hazardous cargo as general freight on roads and railways means that the shipment is transported under federal regulations that do not mandate prenotification or continuous monitoring en route. It is, in short, a regulatory regime designed to expedite commerce. The adequacy of DOT regulation of hazardous materials transport has been criticized by the General Accounting Office, citizen groups, and the Office of Technology Assessment, among others. But even after Congress questioned the adequacy of federal rail regulations, in hearings on an Amtrak-Conrail wreck that killed 16 and injured 197, DOT planned to terminate funding for state rail inspection programs. These programs, according to the GAO, cost a little over $2 million a year, and in the last few years, they accounted for 25 percent of all government railroad inspection personnel.

Until recently, drivers were not required to have a special license or to have demonstrated a safe driving record before they were permitted to haul a hazardous cargo.[16] Drivers of vehicles carrying such materials supposedly received special training, but the only required proof of such training was a certificate that could be issued by the trucking company. The 1986 Commercial Motor Vehicle Safety Act now requires DOT to establish minimum federal standards for the licensing, testing, and qualification of any person operating a commercial motor vehicle carrying a hazardous material. The Act is intended to remedy past abuses, such as drivers evading state regulations by carrying multiple licenses. Through a national registry of commercial vehicle operators, monitoring past violations is intended to become a regular part of driver licensing. The Act also includes provisions aimed at improving the adoption and implementation of state vehicle-inspection programs, and federal regulations governing truck braking systems. However, the specifics of this Act's implementation and the effectiveness of its enforcement are still uncertain.

Provisions found in the Motor Carrier Safety Act do not apply to railroads. Railroad engineers are trained in-house by the rail company. There is no such thing as a state or federal train engineer's license. Train crews seldom receive special training for emergency response or hazardous materials transport. A survey of National Transportation Safety Board accident reports found that employees generally knew little about what they were hauling, or appropriate emergency response measures. The Board has repeatedly criticized the adequacy of railroad emergency response planning. In some accidents, trains were found to lack even a working radio that could be used to summon energy personnel to the scene.

The profusion of hazardous shipments and inadequacy of federal monitoring and enforcement have prompted states and local governments to attempt their own forms of regulation. Such attempts have included requiring special permits, and rail car and vehicle inspections, levying fees on radioactive shipments, requiring that hazardous cargoes bypass city centers, and imposing time-of-day and routing restrictions. In most cases, the DOT has ruled that such regulations are inconsistent with federal regulations designed to protect the flow of interstate commerce. Permits and fees targeted specifically at radioactive cargoes have usually been ruled an unacceptable barrier to interstate commerce. At first DOT even opposed plans by Columbus, Ohio, to erect a sign directing trucks carrying hazardous cargoes to use an interstate beltway around the city, although federal regulations require that such routes be used for interstate shipments where available. The Department of Energy has also joined with DOT to challenge such state and local regulations. In some cases, however, court decisions have been more responsive to public concerns about the health and safety risks. The U.S. Supreme Court, for example, ruled recently that Illinois' levy on shipments of spent reactor fuel was legal because the money was used to fund state monitoring programs and emergency preparedness. Nevertheless, communities contemplating more aggressive oversight of hazardous cargoes—whatever the type of cargo—must be prepared to confront lengthy and costly legal challenges intended to discourage such attempts.

LOCAL COSTS

The transport of hazardous raw materials and the products of weapons industries burden state and local jurisdictions charged with protecting the health and safety of their constituents. Local communities must bear the immediate costs of responding to transport accidents and can recover damages only through legal proceedings. An emergency response to a recent Amtrak-Conrail accident that involved no hazardous materials still cost one local fire department in excess of $50,000. Such costs are burdensome for any community, but especially for the majority of communities in the U.S. that must rely on volunteer fire departments. Nevertheless, transporters assume that an emergency response capability is in place and that it is capable of handling any foreseeable emergency.

Carriers themselves offer little in the way of an emergency response capability. Many rail yards possess virtually no emergency equipment. In its report on the nitric acid spill in Denver, the National Transportation Safety Board found that the railroad company did not have in place an emergency plan for accidents or incidents in its Denver rail yards, and no prior arrangements had been made with Denver public safety organizations for them to respond. Even where available, the Board has found the quality of such plans to be variable, as is their integration with local government plans and capabilities. Thus, the burden of emergency response falls first, and primarily, on local responders.

Large cities can finance police, fire, and public health departments with the sophisticated knowledge and equipment needed to respond to an accident involving hazardous materials, but such capabilities cannot be marshaled if the nature of the military cargo is unknown. Placards identifying the contents of a trailer or boxcar may be destroyed in the accident. Information on shipping papers may be inaccurate or lost. Information on military shipments may be difficult to obtain in a timely manner—as seen in the Denver torpedo spill example. Improper emergency response procedures can exacerbate a serious threat to public health or, at the least, complicate emergency responses. Even the best preparations may be inadequate in unusual or extreme accident situations involving military cargoes, and for most high-consequence but rare accidents there is no base of experience to guide decisions.

Accidents often occur in locations where first responders lack sophisticated training and equipment. Nearly three-quarters of the nation—including the majority of rural areas—are served by part-time or volunteer fire departments. Rural states such as Wyoming can afford to employ only one or two people knowledgeable in radiological health and emergency response. Assurances that DOE's radiological assistance teams or chemical industry experts are available to assist the states are of little comfort when such teams may take hours—perhaps up to a day—to reach an accident site. The impacts of any serious accident in a remote location will be compounded by the likelihood that the accident will be discovered by a person with little understanding of a military cargo or the hazards it presents.

Added to these concerns are questions about who will pay the costs of responding to, and cleaning up after, a transportation accident. Under the Motor Carrier Safety Act, carriers of significant quantities of radioactive materials are required to carry $5 million in insurance. The adequacy of this coverage is questionable, especially if an accident involves a large-scale evacuation and complex decontamination procedures following an accident. However, the Price-Anderson Act covers many types of accidents involving nuclear materials, including those associated with transportation to and from production facilities operated by DOE contractors. Once liability and damages have been established under a state's tort law, the Price-Anderson Act establishes a system for paying damages. (It should be noted that many privately owned facilities apart from nuclear reactors are not covered by the Act.) In cases where the shipper's or receiver's coverage does not include transportation accidents, the Price-Anderson Act provides $160 million for the compensation of victims. Although the Act generally covers the costs of state and local emergency response, including precautionary evacuations, the $160 million transporters' insurance policy does not and if the radioactive material is stolen and later released, the resulting injuries are not covered by the Act. Similarly, most private insurance policies exclude damages caused by war or civil insurrection, which could exclude damages resulting from sabotage. Under existing arrangements, victims of this type of transportation accident would probably never be compensated.

It should be noted that the Price-Anderson Act applies only to radioactive materials. There is no equivalent statute for accidents involving toxic chemicals or many other dangerous components of weapons systems that are transported on U.S. rail lines and highways. Definitions of liability and the sufficiency of private insurance to deal with accidents involving military-related cargoes remain an open question. Liability and compensation for transportation accidents involving hazardous materials, and agencies of the federal government such as the Department of Defense, raise complex questions about the level of protection and the costs that victims of such accidents will be forced to bear.

When considering the local costs of military transport, state and local jurisdictions are often placed between citizens concerned about catastrophic accidents, and federal agencies that have decided that the risk is acceptably low. States may lack the resources to fund the emergency response preparedness programs necessary to assure their citizens that an accident can be managed safely. Western states, for example, point out that they lack the radiation detection equipment necessary to assure the public that packages en route to WIPP are not leaking. When federal agencies do not meet state and local demands for additional equipment, training, and other protective measures, and when state and local governments cannot assure the public that adequate resources are in place, public confidence in all levels of government suffers.

Transportation is one of the more visible targets of citizen protest against the military-industrial sector. Shipments are relatively easy to track, and the techniques for disruption easier than, say, those for disrupting the internal workings of a major weapons facility—as protests at the Rocky Flats plant near Denver have shown. However, local communities must bear the costs of providing police and prosecuting civil disobedience directed at military shipments. This puts local jurisdictions in the situation of having to protect public health and safety, and yet wanting to avoid the costs of citizen protests addressed to those same issues—a contradiction that pushes many jurisdictions into working with federal agencies to assure the secrecy of provocative shipments.

There are many other examples of the hidden costs associated with military cargoes. For years nerve gas was stored at the Rocky Mountain Arsenal, adjacent to Denver's Stapleton International Airport. That gas was later transported to Utah for disposal at the Tooele Army Depot, en route passing through the Salt Lake City metropolitan area. This one example illustrates the health and environmental burdens associated with the disposal of obsolete materials from military sites. As cleanup of DOE facilities is undertaken, inevitably some of the waste from the hundreds of toxic waste sites found on military facilities will eventually be transported to disposal sites. High-level radioactive waste transported from weapons production facilities such as Savannah River may eventually be transported through Las Vegas and numerous other metropolitan areas on its way to a repository in Nevada. The containers used to ship these materials are not totally impermeable. It is assumed that residents along transportation corridors will incure some additional exposure to radiation, but at a level that authorities estimate will fall within the standards of "safe" exposure, provided containers are manufactured, maintained, and handled correctly.

Finally, local responders are concerned about their liability in the event of an accident involving these sorts of materials. Many state and local governments are concerned about a situation in which they may take the wrong course of action, either through misinformation or lack of resources, and later be held liable for the resulting damages.[17] Estimates of damage produced by a transportation accident involving high-level waste are controversial and depend on one's view of likely transportation accidents. Figures cited in a 1980 NRC report, *Transportation of Radionuclides in Urban Environs,* estimated such damages at $1.8 billion for a densely populated urban area, an estimate challenged by many environmental groups as ridiculously low. Although dollar estimates vary, it can be agreed that, at a minimum, military transport confronts local economies with the possibility of evacuations, closed rail lines and highways, and the unknown, long-term effects from exposure to toxic and radiological materials used in the manufacture of weapons.

Estimates of the probability of a transportation accident are beside the point so long as local communities are expected to be able to respond to such an accident. Even if an accident never occurs, local communities must still assume the risks and burdens associated with the everyday transport of materials to weapons industries and facilities. The traffic in military cargoes may have been assimilated into everyday *economic* activity, but it has yet to become the subject of everyday *political* activity.[18] States and local governments are just now coming to terms with the extent of the traffic in hazardous materials. The transport of many military cargoes poses the same problems and questions as the transport of any hazardous cargo, but many military shipments—including those of radioactive materials—are exempt from public scrutiny and environmental controls. Measures viewed as prudent and desirable for commercial shipments are not required for military shipments. The potential for catastrophe posed by the traffic in military cargoes remains an ever-present, yet largely unaddressed, consequence of preparing for war.

TERRORISM AND THE ISSUE OF NATIONAL SECURITY

So far, our discussion has focused on the *outcomes* of federal policies on the movement of military cargoes, which have much to do with the everyday life of the community. The secrecy that surrounds the transport of materiel is predicated upon the strategic importance of these cargoes, and one of the important *inputs* to policy determination is the assumption that weapons are threatened by saboteurs and terrorists.[19]

In the mid-1970s, the threat of criminal action against nuclear facilities in the United States could have been dismissed as of marginal importance, because neither terrorists nor organized crime were seen as possessing the capabilities to obtain or utilize nuclear materials.[20] Within three years, the proliferation of terrorist activities throughout the globe suggested that the putative threat to the nation's nuclear facilities should be taken seriously.[21] Since the early 1980s, several dozen "incidents" have taken place in and around nuclear plants, but published reports appear to have focused primarily on the profiles of terrorist groups rather than on the vulnerability of different facilities.[22] There is little published material on the movements of nuclear products, although one 1977 news report pointed to the existence of "plutonium police"—the Transportation Safeguards Division, organized by the Energy Research and Development Administration (now DOE). When this group was created in 1974, it employed more than 125 courier-guards and cost over $12 million per year.[23] Its personnel were under orders not to permit police to examine their vehicles, or even to leave their cabs if persons without security clearance were present.

It is difficult to assess the risks posed by terrorist activity. On the one hand, the NRC revised early risk estimates (applied to a commercial waste shipment) of 1,800 latent cancer fatalities following an attack, downward to 14 fatalities.[24] In consequence, the commission proposed removing armed guards from such shipments. On the other hand, many shipments of materiel are not accompanied by professional security personnel, and we now consider these in light of the complexity of the military transportation network in the United States. Under even the most conservative definitions,[25] at least 300 federal weapons-related facilities throughout the country, including bases, exchange chemicals, finished products, conventional weapons, and obsolete hardware. Simple mathematics suggests something like 90,000 possible linkages among individual sites.[26] Even if there were only one shipment between each facility every year, that would still constitute a very large number of shipments to oversee. We may reasonably assume that terrorists of any stripe would be capable of following trucks from the facility of origin and keeping them in sight until the time was ripe for attack. The chance of success would often be high, insofar as the vast number of trucks on the highway system makes careful oversight by law enforcers impossible.

We cannot anticipate precisely the nature of the materials that could be obtained in such an attack, or the uses to which they could be put. Although the theft of a nuclear device might seem the most alarming possibility, even sophisticated terrorists might be hard-pressed to arm a nuclear warhead, which cannot be detonated merely by hitting it with a wrench. However, the threat of environmental contamination by releasing the contents of a nuclear weapons package or a hazardous chemical could be realized much more easily, and could be used in a ransom or hostage situation. Many military-related cargoes apart from actual weapons may seem more vulnerable and thus have a greater attraction for the terrorist.

So far, our discussion has concentrated upon the transport of military cargoes within the United States, but ours is by no means the only country burdened with such risks. In Yorkshire, England, for example, local communities claimed recently that shipments of nuclear warheads were being made along public roads, with trucks transporting the materiel from Berkshire, in the south of England, to the Coulport submarine base in

Scotland. Local politicians claimed that they were given no official information concerning the shipments, and had no knowledge of the necessary emergency preparedness requirements.[27] Not only do most nations move military cargoes about within their territories, many nations also move hardware through intermediate countries en route to third parties. Given the large number of military bases established worldwide by nations such as the United States, the Soviet Union, France, and Britain, and the sale of military material to other countries, we can assume a continual movement of new and obsolete materiel across borders and oceans. The trade in arms all over the globe represents one of the largest categories of international trade. *The New State of the World Atlas* for 1986, for instance, indicates that very few nations are not involved in the arms trade in some form.[28] Thus, many of the issues of transportation safety that we have raised here are of international significance.

CONCLUSIONS

It would be a mistake to relegate the issue of transportation to the sidelines in any analysis of the environmental consequences of preparing for war. First, the transport of military cargoes is by definition hazardous; modern history is literally littered with such accidents. Indeed, one of the largest single industrial accidents in North America occurred when a munitions ship blew up in Halifax, Nova Scotia, in 1917, killing 1,500 people and destroying much of the city. Second, there are long-term risks, as when nuclear materials emit radiation while being transported. The levels of radiation involved may be unimportant, but given the secrecy that surrounds weapons transport, cargoes are not open to public scrutiny and no certainty can be attached to any assertion of safety. Third, the movement of military cargoes involves numerous costs. States and localities in the United States—and indeed communities throughout the world—must train and prepare for catastrophe. Not every locale will use its training, but emergency preparedness must be organized nonetheless. When problems do occur, the response must be adequate in order to avoid worsening the consequences. Moreover, most accidents necessitate evacuation, and restitution afterwards, which can be costly.

Fourth, transportation increases the probability of a successful terrorist attack. As long as products are stockpiled in well-guarded military spaces, the likelihood of attack is remote. The chances of a successful raid against a military convoy are probably low, but the odds are much higher when single vehicles operated by commercial truckers are taken into account. Fifth, the need to organize and control many thousands of shipments annually fosters secrecy, discourages public awareness, and creates a climate in which public protest is stifled. Communities are often ignorant of what passes through their territories, and this ignorance extends to state governments as well. Federal preemption represents another instance of what some regard as the growing power of the state apparatus, which can wrap its actions in a cloak of secrecy and misinformation.[29] Sixth, the movement of materiel is the reification of a world weapons economy. The weapons trade has functioned alongside transnational banking and international borrowing to create a class of nations bound through indebtedness to more affluent neighbors. A spiral of obsolescence and further spending is creating an arms race among less-developed countries that may not destroy the globe but is certainly increasing the death toll from local conflicts.

In this chapter, we have attempted to consider the profusion of cargoes and issues that come under the heading of military transport. In the end our perceptions—like those of anyone living near a rail line or highway corridor through which cargoes of toxic terror travel—are filtered through the lens of the locality. We draw on the book *White Noise*, by Don DeLillo, for the last word on the implications of a transportation accident and evacuation.

> A few minutes later, back on the road, we saw a remarkable and startling sight. It appeared in the sky ahead of us and to the left, prompting us to lower ourselves in our seats, bend our heads for a clearer view, exclaim to each other in half finished phrases. It was the black billowing cloud, the airborne toxic event, lighted by the clear beams of seven army helicopters. They were tracking its windborne movement, keeping it in view. In every car, heads shifted, drivers blew their horns to alert others, faces appeared in side windows, expressions set in outlandish wonderment.
>
> The enormous dark mass moved like some death ship in a Norse legend, escorted across the night by armored creatures with spiral wings. We weren't sure how to react. It was a terrible thing to see, so close, so low, packed with chlorides, benzines, phenols, hydrocarbons, or whatever the precise toxic content. But it was also spectacular, part of the grandness of a sweeping event, like the vivid scene in the switching yard or the people trudging across the snowy overpass with children, food, belongings, a tragic army of the dispossessed. Our fear was accompanied by a sense of awe that bordered on the religious. It is surely possible to be awed by the thing that

threatens your life, to see it as a cosmic force, so much larger than yourself, more powerful, created by elemental and willful rhythms. This was a death made in the laboratory, defined and measurable, but we thought of it at the time in a simple and primitive way, as some seasonal perversity of the earth, like a flood or tornado, something not subject to our control.[30]

NOTES

1. This is argued by several authors in A. Blowers and D. Pepper, eds., *Nuclear Power in Crisis,* Beckenham, Croom Helm, 1986. A more radical interpretation would argue that the nuclear state is not a separate entity but is only a part of the larger state apparatus. See A.M. Kirby, "The Political Implications of High-Level Nuclear Waste Transport," *Government and Policy,* (1988) 6(3), 311-22.
2. We should not lose sight of the ways in which language can obscure issues of risk and safety. In researching this chapter, we found, for example, a newspaper article entitled, "Army Says Germ Warfare Lab Would Be Safe." (*New York Times,* Feb. 1, 1988.)
3. In Great Britain, official calculations equate the risk with that of smoking one cigarette in a lifetime: see Kirby (1988).
4. The Office of Technology Assessment (OTA) concluded that costs of accidents are underestimated, noting that only 10 to 25 percent of all hazardous materials accidents may be reported. The reliability of accident data and OTA's conclusions are controversial. Less than 3 percent of the reported accidents have involved radioactive materials, but the proportion of shipments involving radioactive materials has also been proportionately small. U.S. Office of Technology Assessment, 1986. *Transportation of Hazardous Materials,* U.S. Government Printing Office, Washington D.C.
5. See for example J.H. Johnson and D.J. Zeigler, "Evacuation planning for technological hazards," *Cities,* May 1986, 148-56.
6. There will be some instances, particularly in the western states, where very large counties extend beyond this distance threshold, but such jurisdictions typically have very low populations and thus do not greatly bias our calculations. For details of routing, see U.S. Nuclear Regulatory Commission, *Public Information Circular for Shipments of Irradiated Reactor Fuel,* NUREG-0725, Washington D.C., 1984.

7. U.S. Office of Technology Assessment, *Transportation of Hazardous Materials.* U.S. Government Printing Office, Washington D.C., 1986. The OTA report incorrectly identified the accident as taking place on a Sunday. It in fact occurred before rush hour on a Wednesday (*Denver Post,* Aug. 2, 1984).
8. As early as 1979, some 78 communities in the United States, including large metropolitan areas like New York City, had passed ordinances attempting to limit the transport of nuclear and/or hazardous materials.
9. See R.M. Bernard and B. Rice, *Sunbelt Cities,* Austin, University of Texas Press, 1984.
10. The 45 trucks in the U.S. Department of Energy's own nuclear weapons fleet alone travel an estimated 4 million miles a year. Agency traffic records show that 150 accidents have occurred since 1975, although the agency claims that only 11 involved major damage to the tractor trailer. DOE claims that no spills of radioactive material from these convoys have occurred in more than a decade of transport, although no outside source can confirm this.
11. U.S. General Accounting Office, *Nuclear Health and Safety: DOE Needs to Take Further Actions to Ensure Safe Transportation of Radioactive Materials,* GAO/RCED-88-195, Washington, D.C., 1988.
12. In 1988, the U.S. General Accounting Office noted that despite a Westinghouse study that found potentially serious flaws in DOE shipping containers, the agency continued to use them to ship high-level radioactive materials. U.S. General Accounting Office, 1988, op cit.
13. Various reports in the *Washington Post,* 1983 to 1988.
14. The shadowing of military convoys has become common among antinuclear activists. Groups such as *Nukewatch,* based in Wisconsin, are committed to the curtailment of what they describe as H-bomb convoys. In the case of other hazardous materials, states may obtain information on the volume of dangerous materials flowing through their jurisdiction, but prenotification for nonradioactive hazardous cargoes, such as explosives and toxic chemicals, is virtually non-existent. Thus, many dangerous shipments continue to move through communities without citizens' knowledge.
15. *New York Times,* Oct. 11, 1987.

16. A 100-hour survey of truck movements, conducted by Colorado Ports of Entry Division in 1985, found that 11.9 percent of those drivers of trucks carrying hazardous cargoes had received *no* driver training in the previous two years. The Commercial Motor Vehicle Safety Act of 1986 requires the DOT to establish minimum standards for the testing and licensing of all drivers of commercial vehicles.

17. For an overview on the complexities of liability, see O. Brown, *Insurance-Indemnity Coverage for Nuclear Transportation*, Sandia National Laboratory Contractor Report, SAND85-7260, 1985.

18. In addition, commerce may itself be one way for agencies to avoid facing political opposition over the materials associated with military programs. Nuclear materials moved across several national boundaries can become "lost in transit" before being diverted to weapons production. A recent case involving mislabeled shipments of nuclear materials between Belgium and West Germany by Transnuklear raised suspicions that the materials were destined ultimately for weapons programs in Pakistan and Libya.

19. See P. Leventhal and M. Hoenig, "The Hidden Dangers: Risks of Nuclear Terrorism," *Terrorism*, 1988, 10: 1-22.

20. B.M. Jenkins, *The Potential for Nuclear Terrorism*, Rand Corporation, Santa Monica, CA, 1977, Paper #5876A.

21. See G. Bass et al., *Motivations and Possible Actions of Potential Criminal Adversaries of U.S. Nuclear Programs*, Rand Corporation, Santa Monica CA, 1980, Paper #R-2554-SL.

22. There is some evidence that security at civilian nuclear power plants was increased, for it is clear that such plants have been vulnerable. It is estimated that "there are more than 9600 lbs. of enriched uranium and plutonium unaccounted for in the U.S." For example: J. Denton, "International Terrorism—The Nuclear Dimension." *Terrorism*, 1987, 9(2): 113-23.

23. *San Francisco Bay Guardian*, Aug. 4, 1977. More recent reports indicate that the Transportation Safeguards Division of the DOE now possesses 45 tractors, 34 trailers, 49 escort vehicles, and 160 drivers.

24. U.S. Office of Technology Assessment, 1986, op cit., p. 106.

25. These calculations exclude post offices, coast guard stations, landfills, and agency offices without obvious weapons capabilities. Source: Federal Facilities Docket, Sept. 21, 1987.

26. Combinatorial mathematics shows that with n nodes, there are {n (n-1)} possible links in the system. That is, with 300 nodes (facilities), there are 300 x 299 links in the circulation system.
27. *Yorkshire Post*, England, July 27, 1987.
28. M. Kidron and R. Segal, *The New State of the World Atlas*, London, Pluto Press, 1986.
29. See, for instance, A. Giddens, *The Nation State and Violence*. Cambridge, Polity Press, 1985.
30. Don DeLillo, *White Noise*. New York: Viking Penguin Books, 1985, p. 127.

6

Venting, Shaking, and Zapping: The Environmental Effects of Nuclear Weapons Tests

JAY B. SORENSON

TODAY, MORE THAN THIRTY YEARS AFTER NUCLEAR WEAPONS TESTS ON THE Marshall Islands and almost fifty years since Hiroshima and Nagasaki, we can make several observations:

First, the history of the tests tells a troubling story of important adverse or uncertain environmental effects, of government negligence, coverups, disinformation; and of failure to protect civilians, employees, and servicemen. In this respect, it tells the story of abuse of power and responsibility, a flawed accountability system, and a failed Congressional oversight process.

The adverse environmental effects are partially explained by scientific uncertainty, but are more fundamentally attributable to a mindset so focused on secrecy and national security that defense is divorced from environmental protection: safety and health concerns are ignored, and environmental costs are not factored into decision cycles; the attendant result is not only environmental harm but an erroneous impression that nuclear weapons are "cheap" in comparison with other weapons systems.

Second, while many changes are occurring in Eastern Europe and the Soviet Union that are overtaking the Department of Energy (DOE), it is

---

The author wishes to express his appreciation to Tracy Lucas-Schouard for her assistance in this study.

still striving to modernize its capacity to produce more as well as new weapons in the future. Its future modernization plan is based on the assumption that nuclear weapons will remain the principal element in the American deterrent system for probably the next 50 years and that preparations have to be made to meet several possible contingencies: 1) nuclear weapons will have to fulfill new lethality functions under the numerical weapons reduction requirements of the START regime and other future arms control agreements; or 2) the current stockpile may have to be expanded;[1] or, finally, 3) arms control and detente efforts could fail.

DOE is using the need for state-of-the-art environmental compliance at its production facilities as a partial justification for building new facilities with expanded capacity. While its materials production reactors are temporarily shut down for safety reasons, the current nuclear weapons production program had been built up to a 20-year peak capacity and was capable of producing 2,000 warheads annually. An even greater capacity is to be provided for with its modernization plans.

Third, the new program indicates that the Department of Energy anticipates increased weapons tests in the future. It is seeking technical-fix advances in miniaturization, mobility, and accuracy, greater penetration and lethality, and star-wars weapons to secure unilateral defense objectives. The problem is that continued testing could ratchet-up the arms threat, and thus either weaken present efforts at confidence building through arms control regimes or result in a dramatic escalation of the arms race.

This chapter traces the environmental effects of nuclear weapons testing, but, as will be discussed below, it proposes an interlocking, self-enforcing system of controls that obviates the need for DOE's modernization, expansion, and testing program. Called a Verification Triad, for want of a better name, it incorporates START and test ban provisions, but it includes a materials production ban. For its elemental verification measures, it is modeled on the Intermediate Nuclear Forces Treaty negotiated by Reagan and Gorbachev in 1988. It is proposed as a measure to reinforce confidence in defense at lower levels of threat, support a more stable and cooperative relationship, and assure a better protected environment.

ATMOSPHERIC TESTS

On March 1, 1954, Aikichi Kuboyama, one of 23 Japanese fishermen of the *Lucky Dragon* exposed to radioactive fallout from the BRAVO atmospheric test of a hydrogen bomb exploded at Bikini Atoll, in the

Marshall Islands, died in a Tokyo hospital. Days before, the Atomic Energy Commission (AEC), precursor to DOE, announced that 28 of its observers and 236 natives on islands scores of miles from the test site had been showered with a radioactive "fallout."[2]

Intended as a "routine atomic test," the BRAVO thermonuclear detonation had apparently gotten "out of control." Government reports indicated a yield equivalent to 15 megatons of TNT and effects "three or four times" larger than expected. BRAVO's radius of total destruction was supposed to be 10 miles or more; its initial radioactive fallout contaminated an area up to 300 miles downwind from the explosion and 40 miles wide.[3] BRAVO's immediate fallout burden fell on the *Lucky Dragon*, carrying a harvest of tuna and shark in an open hold. Caught 80 miles from the explosion, the fishing vessel was showered with a gritty white ash of particles about 90 minutes after the explosion. The fallout blistered the skin of the 23 fishermen and rendered the fish radioactive.[4] The total external radiation dose to the fishermen was estimated to be 170 to 600 rads. The lethal dose in humans—causing a 50 percent chance of death—is estimated to be 450 rads. Four to six hours after the detonation, prevailing easterly winds blanketed Rongelap Island, 120 miles away, with a white, powdery radioactive material. People were exposed to approximately 175 rads of whole-body radiation. The radioactive dust took about 22 hours to reach Utrik Island, 300 miles east of Bikini Atoll. People there were exposed to about 14 rads. Both islands were hurriedly evacuated several days later.

The tragic mishap of the BRAVO test, more clearly than Hiroshima and Nagasaki, pointed up for the first time the importance of fallout as a significant source of radioactive contamination.

The United States tried to put the best face on BRAVO's unintended consequences, which triggered a worldwide demand for an international ban on nuclear weapons tests. AEC Chairman Lewis L. Strauss witnessed the test and quickly flew to Washington to allay a growing public concern. Standing beside President Eisenhower at a press conference, Admiral Strauss, in an early effort at disinformation, stated that the reports had been "exaggerated." The "yield" had been about twice that estimated: "a margin of error not incompatible with a totally new weapon." No large island had been blown up—only small sandbars. The *Lucky Dragon* had been the victim of an unfortunate wind shift. Contradicting the AEC's earlier report, he said the fishermen's wounds were probably due not to radioactivity but to chemical burns caused by lime particles from blasted coral. It was not possible to know, he told the nation, since the Japanese had not allowed U.S. doctors to examine the victims, but the United States was prepared to indemnify them. He noted limited adverse effects: there was no widespread contamination of fish in the Pacific. Admiral Strauss said he had visited the injured island natives and found them apparently "well and happy." As for the tests, they had been successful, and the United States could be assured it was keeping up with the Soviet Union.[5] This meant that nuclear weapons were to be key to our approach to competition with Moscow.

A-bomb tests had preceded the hydrogen device tests. Operation Crossroads began at Bikini in 1946 and at Eniwetok in 1948. Code-named Able and Baker, these tests were designed to determine A-bomb effects on the navy and to learn how to make smaller bombs. Forty-two thousand servicemen and a target fleet of some 70 anchored ships participated in Able and Baker. Following the tests, radioactivity in Bikini lagoon remained far greater in intensity and persisted longer than had been anticipated. Nine thousand largely unprotected men boarded the contaminated target vessels. Dr. David Bradley, in charge of the ship safety survey laboratory and responsible for the analysis of 3,000 samples of urine, requested a manual for "some guidelines or some standards." The colonel in charge said, "You cannot have it." Dr. Bradley asked, "Why not? I've got to. How can I operate?" The colonel replied, "It is secret. It is classified. We cannot leave that information out here. Do the best you can."[6] A 1985 General Accounting Office report found negligence: "comprehensive personnel decontamination procedures were lacking or not followed"; "no estimates were made for radiation exposure through ingestion or open wounds"; and "the estimate for inhalation may be in error by a factor of 5 or 10."[7]

The United States conducted 66 tests in the Marshall Islands from 1946 to 1958: 43 on Eniwetok Atoll and 23 on Bikini Atoll. The environmental damage was extensive; 12 of 66 islands were reported destroyed or rendered uninhabitable because of fallout. All were covered with some 240 long-lived radionuclides, including radioactive isotopes of trace elements necessary for human life, which concentrate in food chains and accumulate in the human body.

By 1967, AEC scientific surveys established that Bikini's radioactivity had declined to levels acceptable for human habitation. In 1968, President Johnson agreed to return the Bikinians to the Bikini-Enyu Island complex, but with special precautions to be observed in the use of birds, turtles, and certain plants on other islands as food sources. Beginning in 1969, the island was cleared of nuclear test debris and a heavy growth of post-test scrub vegetation. As a radiological precaution, the top two inches of

Bikini's topsoil were removed before the planting of pandanus trees. In 1970, coconut and food-crop planting was begun. Several dozen people were repatriated for a few years. However, additional scientific studies indicated that life on the island was dangerous, and the people were forced to leave.

To date, $90 million has been spent for cleanup. Thousands of samples of soil and fruit indicate that radioactive isotopes such as strontium 90 are still in the topsoil and that there is a higher concentration of cesium 137 in coconuts than in the topsoil. The current assessment is that Bikinian radiation exposure would be low level and long term rather than acute. Radioactive uptake in trees and fruit apparently can be blocked with potassium used as a fertilizer, but radiation would remain in the soil unless the topsoil is scraped away. Bikinians seem to prefer soil excavation to the use of potassium. William Robinson, a Lawrence Livermore Laboratory scientist responsible for testing different radiation control and cleanup methods, fears excavation means the island will be destroyed to save it.

Because information on radioactive contamination in other countries is scarce, this chapter focuses on U.S. tests. The result may be the appearance of a distorted view of the environmental consequences of U.S. weapons testing; there is good reason to believe the effects are worse elsewhere.

Weapons Effects

Much is still to be learned about direct and indirect injury and the lethal effects of nuclear weapons to humans, other organisms, and the environment. Lethality is a function of blast, thermal radiation, initial and residual ionizing radiation, and of the distance from the hypocenter of a detonation to the outer boundary. A 50 percent lethality is indicated when the number of survivors equals the number of fatalities in an exposed population.[8] Standard relationships for direct effects are based on observations from Hiroshima and Nagasaki and subsequent weapons tests.

Much physical damage is due to an intense shock wave created in the atmosphere—also referred to as "blast." The intensity of this shock wave is measured at its peak in units of pounds per square inch (psi). This blast intensity (overpressure) decreases as the shock wave travels away from the epicenter of the explosion. The radial distance at which the overpressure drops to 5 psi is considered to be the distance over which the physical destruction of most buildings is complete. For a one-megaton explosion, the 5 psi level occurs at a distance of 4.4 miles and covers an area of 60 square miles. People may be killed directly by blast. The lethal dose for half of an exposed population is about 12 psi. Injury occurs at 2 psi. However, collapsing buildings and flying projectiles produce injury and death at much greater distances.

Heat and fire cause great damage and many deaths over short and longer ranges. Enormous energy is released and forms a thermal pulse that travels radially at the speed of light, causing severe burn injuries to every living thing in the direct line of sight of the fireball.[9] The thermal pulse can ignite fuels and combustible materials at considerable distances from the blast epicenter. Blast wave winds extinguish many of the initially set fires, but new fires start. These can create firestorms and contribute to chemical pollution and damage to the ozone layer, as well as leading to nuclear winter and climate change (See Chapter 7).

Initial ionizing radiation, emitted directly from a nuclear detonation, consists primarily of gamma rays and fast neutrons. However, the lethal area for direct ionizing radiation is smaller than the lethal areas for blast and thermal radiation. Since there is a time lag of hours, or even weeks, for death due to exposure by direct radiation, most deaths of people, animals, and vegetation in the lethal area will be caused principally by heat and blast. This is not the case for an enhanced radiation weapon, the neutron bomb, which kills primarily with radiation and produces minimal heat and blast effects, leaving buildings, military equipment, and other structures standing.[10]

Local and regional radiation exposure is difficult to predict. It depends on the weapon type, yield, burst height, weapon material, and the direction and intensity of winds and rain at all altitudes, not just near the ground. Individual exposure depends on the local terrain, time spent outdoors, type of building construction, and adequacy of shelter. Direct radiation fallout exposures can exceed thousands of rems (roentgen equivalents in man) and, if from large surface bursts, can extend hundreds of kilometers downwind.

Global fallout consists of fine radioactive dust or aerosols. Weapon debris and other material is vaporized by the high temperatures and pressures of fireballs and recondensed in the atmosphere to form fine particles. A large portion of these particles are sufficiently massive to be deposited within 24 hours or so after the detonation. The particles lay a carpet of intense radioactive fallout in an elongated pattern immediately downwind from the site of the explosion. Most of the finer radioactive particles and aerosols stay suspended in the troposphere for weeks to months, and in the stratosphere for month to years, before returning to the Earth's surface. Recent calculations[11] suggest that global fallout in the Northern Hemisphere is on the order of a few tens of rems, levels likely to damage crops and livestock and create a legacy of birth defects.

Nuclear weapons tests during 1961 and 1962 are also found to have had a major impact on the upper atmosphere. The important factor seems to have been the absorption of light by oxides of nitrogen created in the fireball.[12]

Cancer Incidence

Unlike military personnel at Bikini and Nevada and civilians at Nevada, the Rongelap islanders are one of few groups studied for fallout effects. Within 48 hours, two-thirds of the exposed people developed nausea, a loss of appetite, and in some cases diarrhea and vomiting. Over the next one to two weeks, 90 percent developed beta radiation burns of the skin; half developed partial hair loss. White blood cell and platelet counts remained abnormal for one to two years. Thirty-four percent of the Rongelap and about 10 percent of the Utrik population developed thyroid tumors, believed to have been caused primarily by short-lived radioactive iodine. Chromosomal aberrations and increases in miscarriages and stillbirths were noted during the first two years after exposure. Transient alterations in immune response occurred in this group, and one death resulted from leukemia. One-year-old children were estimated to have received larger doses—up to 2,000 or more rads. Two young boys suffered growth retardation caused by thyroid gland damage.[13] About 77 percent of Rongelap children under ten years old at blast time required surgery later for benign and malignant thyroid nodules.[14]

Worldwide effects of atmospheric tests are uncertain. In 1979, Robert E. Alexander, a Nuclear Regulatory Commission staff member, guessed that the long-term worldwide effect may be 29,000 to 72,000 deaths from whole-body exposure and 168,000 cases of genetic change and possible birth defects for all generations.[15]

Cancer Standards: Hiroshima, Nagasaki, and Little Boy

Worldwide radiation standards are based on studies of Japanese survivors of Hiroshima and Nagasaki, known as the Hibakusha. Standards are fixed according to calculations of radiation doses and the records of subsequent leukemias, cancers, or birth defects. The studies assume exposures to direct radiation, with only minimal exposure to fallout. After leukemia deaths peaked in the first five years, a persistent increase in other cancer deaths occurred. In 27 to 32 years after the bombs, the excess cancer death rate had increased by a factor of 2.4, caused by cancers of the esophagus, stomach, colon, lung, breast, and urinary tract, as well as lymphoma and multiple myeloma. Cancer of the thyroid gland and benign tumors also increased.[16] Until recently, many scientists were surprised to see little evidence of a significant increase in birth defects among Hibakusha children. There is, however, evidence of damaged chromosomes in the cells of parents and some children.

Numerous uncertainties about radiation and the exact yields of the two bombs have contributed to a debate about radiation safety standards. New evidence indicates that gamma radiation exposure was underestimated, and neutron radiation exposure was overestimated. A new six-year U.S.–Japanese study indicates that the risk of contracting cancer from gamma radiation is at least twice as great as previously thought.

To provide an assessment base independent of the health effects on survivors, a replica of the Hiroshima bomb, Little Boy, was built at Los Alamos National Laboratory.[16] The 1980–83 research project, directed by Richard Malenfant, was intended to determine exactly what was in Little Boy and the precise radiation dose it delivered. The replica, built to exact original specifications and with enough uranium to sustain a controlled reaction without an explosion, was tested at Los Alamos, with radiation measured up to a half-mile away. The bomb and radiation reconstruction, superimposed on the climatic conditions at the time Little Boy was dropped, has led to new findings. Neutrons from the Hiroshima bomb were contained by several factors: the geometric configuration of the uranium source, the spherical shape of the bomb and its massive steel casing, and the atmospheric humidity in Hiroshima on the day the bomb was dropped. More gamma rays were released than previously estimated; it is now thought that gamma rays cause more health problems than do neutrons.

The tests also cleared up questions of Little Boy's yield. It was 16 kilotons, less than the 20 declared by Truman and more than subsequent 12-kiloton estimates.

Malenfant's findings are consistent with those of the recent U.S.–Japanese study. The U.S.–Japanese team has proposed far more stringent standards for public and occupational protection against gamma radiation. However, Malenfant, joined by James Thompson, a health protection specialist at the Lovelace Inhalation Toxicology Research Institute on

Kirtland Air Force Base in New Mexico, does not believe the "evidence so far is so overwhelming that it warrants a change" in standards.[18] The International Committee on Radiological Protection has deferred a decision on the new recommendations for at least two more years.

Nevada Air Burst Tests

President Harry Truman created the 1,350-square-mile Nevada Test Site (NTS) in December 1950. Sixty-five miles northwest of Las Vegas, it is an isolated area larger than the state of Rhode Island. Fewer than 2,000 people, mostly ranchers and miners, live downwind of it. The closest towns downwind are St. George and Cedar City, both in Utah. NTS wildlife is rich and varied: foxes, kangaroo rats and other small mammals, mule deer, wild horses, centipedes, lizards, rattlesnakes, and bats. More than 30 species of birds are found on the site, including robins, golden eagles, hawks, quail, and chukars. No hunting is permitted on the site. Thunderstorms and dust devils are common at the site, and winds often reach 60 miles an hour. Presumably because of this atmospheric turbulence, deer and sheep 40 miles away have increased iodine 131 in their thyroid glands.[19]

Weapons testing began in January 1951. Some 250,000 military personnel were involved in at least one "Ground Zero" test between 1951 and 1962. Many crouched in trenches just a few hundred yards from the hypocenter. One objective was to observe the effects on the nation's "nuclear warriors." In one exercise, 1,200 paratroopers set up battle positions on the test range, withdrew from the explosion, and then returned within minutes for lessons in decontaminating exposed equipment. As with Marshall Islands military personnel, no retrospective medical evaluations have ever been performed.

The first detonation, code-named "Able," shook the Nevada desert on January 27, 1951. The dirtiest one was during the spring of 1953. A big shot, it partly vaporized the detonation towers. The dust and metal swept up in the fireball eventually rained down as fallout. A DuPont Company photographic paper factory in Rochester, New York closed down after one test; its paper was being ruined by exposure to radioactive dust particles blown 2,000 miles from the test site and caught in the plant's air-conditioning system.[20] In 1955, Apple 2 was shot over a small "city" erected to study urban destruction and civil-defense measures. It left a mass of twisted bridges and burnt-out houses. By the time the 1963 Test

Ban Treaty halted surface shots, more than 183 had been conducted by the United States.

Offsite Civilian Effects

Tests were conducted when winds were not blowing toward Las Vegas or Los Angeles, but at least 87 of 121 atmospheric tests between 1951 and 1958 caused offsite contamination, and 28 spread blankets of radioactive fallout over Utah. The July 14, 1962, "Small Boy" test of about 1 kt was detonated on a tower and produced a "hot spot" of 100 rad/hr in the Orem, Utah area, over 300 miles downwind from ground zero. One hundred-rad readings were also found in areas between St. George and Cedar City, Utah, about 140 miles downwind from ground zero.[21] Test-site manager James E. Reeves reassured his neighbors at the time: "I want you to know that each shot is justified by national and international need and that none will be fired unless there is adequate assurance of public safety."[22]

Utah has the lowest cancer incidence of any state. Yet the cancer rates for nearby offsite areas, or "downwinders," appears relatively high. To evaluate offsite exposure discrepancies, a DOE Dose Assessment Advisory Group worked seven years to recalculate fallout dosages. The dosimetry specialists focused on internal exposures to radioactive isotopes ingested in food and water or inhaled and stored in the body, and on external exposures to radioactive gases and particulates in the air that fell on the skin or came from contaminated soil. They also examined a worst-case scenario for children, the population most vulnerable to radiation. They tested radioactive soil samples and extrapolated backward in time to calculate the original radiation that had fallen in the area.[23]

Average internal exposures to the thyroid gland of 60 rads were found for people in Washington County in the spring of 1953. In a large population, one or two cancers were possible. The number of infants in the area at the time of the tests was probably no more than 20, too few for a statistically significant epidemiological study. DOE's spokesperson, Barbara Yerg, noted studies of dairy "milk sheds" in the community; they showed that regional dairies did not use milk exclusively from cows in contaminated areas, so the worst-case and most dangerous infant exposures did not occur.

Average external exposure was estimated at 5 rads, which is about equal to a lifetime of exposure to background radiation. But all exposures are lifetime cumulative, and 5 rads from fallout is an additional exposure above that experienced from background.

DOE's Dose Assessment Group's findings are at variance with the National Cancer Institute's (NCI) and other studies on leukemia in Utah. NCI's data indicate leukemia rates recorded downwind of NTS from 1950 to 1969 were 143 percent of the state average.[24] Dr. Joseph L. Lyon, codirector of the Utah Cancer Registry, concluded in 1979, after months of research, that childhood leukemia deaths in high-fallout counties of Utah north and east of the test site amounted to 2.5 times the national average.[25] Fifteen to 20 more children under the age of 15 died of leukemia than would have been expected. These excess deaths began appearing in 1959, six years after the tests called Upshot Knothole, known to be the dirtiest series. That six-year gap is consistent with the latency period of leukemia. The death rate peaked in 1967, four years after aboveground testing was halted, then fell to the pretesting level. Some experts considered the odds against all this happening naturally to be 100 to 1.[26]

Two other studies in 1965 and 1982 estimated that the dosage delivered by radioactive iodine to the thyroids of children in St. George, Utah who were under the age of five in 1953, varied between 500 and 2,500 rads. These doses are of the same order of magnitude as those sustained by the Rongelap children.[27]

DOE's dosimetry group also discounted the danger of possible "hot spots" of fallout.[28] Yet reporter Joseph Bauman writes that in 1953, rancher Kern Bulloch of Cedar City, with his brother Mac, was trailing 2,000 sheep back from winter range near the Nevada Test Site. " 'We were out in what they call Tickaboo Valley, and this plane came over and dropped a bomb, and I was looking right at it,' he told me. 'It was just a big ball of fire. Then this big cloud came out. The fallout came right over us. Me and another fellow were in the camp. It drifted over the sheep and over this Lincoln Mine.' "

Later that morning, AEC officials in a Jeep warned the Bullochs they were in "a hell of a hot spot." Bulloch recalled that soon after the flock arrived in Cedar City, "they just started dying like flies. We figured they had lesions in their mouths and lungs. . . . The fallout had settled on the vegetation and they ate it."

Harold Knapp, a health physicist, reported in 1982 that fetal lambs had received fatal doses of 20,000 to 40,000 rads to the thyroid gland, and sheep had received 1,500 to 6,000 rads to the gastrointestinal tract, where the external doses were estimated to be only 4 rads.[29] Of 11,000 sheep and lambs near NTS, 4,300 died in 1953. These were severe and unrecoverable losses for many ranchers.

At first, federal veterinarians told the ranchers they were finding strong evidence of fallout. But later, when the ranchers sought compensation, the AEC denied everything. At a 1955 trial, AEC expert witnesses testified that their investigation indicated the sheep had died of malnutrition and poor range conditions. Judge A. Sherman Christensen found in favor of the AEC.

Twenty-six years later the case was reopened. New government statements indicated that the AEC reports had been misleading and that "the government knowingly disregarded and suppressed evidence correlating the deaths of the sheep to exposure to radioactive fallout."[30] The semi-retired Christensen, rehearing the case, ruled that the federal government had "perpetrated a fraud upon the court." Federal attorneys appealed to the Tenth Circuit Court of Appeals in Denver and won. The U.S. Supreme Court decided in January 1986, by a 5-3 vote, not to hear the ranchers' case.

In 1979, a number of downwinder cancer victims filed suit, charging federal negligence as the cause of their cancer. In 1982, Stewart Udall, former Secretary of Interior, led a team of plaintiffs' lawyers, representing 1,192 survivors or victims of cancer.

On May 10, 1984, Judge Bruce S. Jenkins ruled: "the court finds that the defendant [the federal government] failed to adequately warn the plantiffs . . . of known or foreseeable long-range biological consequences" of nuclear weapons tests. He wrote, "at no time during the period 1951 through 1962 did the off-site radiation safety program make any concerted effort to directly monitor and record internal contamination or dosage in off-site residents on a comprehensive person-specific basis."[31]

He also found that there was no national policy that dictated conducting tests in an unnecessarily dangerous way.

The government appealed. Three years later, a Tenth Circuit three-judge panel unanimously overturned Judge Jenkins on the grounds that testing was conducted within the discretionary powers delegated to the AEC; it was thus immune from liability for negligence.[32]

The case went to the U.S. Supreme Court. In January 1988, in a sovereign immunity decision rooted in English common law—"The King can do no wrong"—the Justices ruled that the government was not responsible for damage done by the testing, even if it acted negligently.[33]

U.S. Atmospheric Tests' Environmental Effects

A succinct summary[34] of environmental effects from atmospheric tests in Nevada and the South Pacific shows that growth abnormalities were observed in fresh vegetation in both areas, reflecting a possible influence of ionizing radiation on rapidly dividing cells. At both sites, long-lived plants with protected perennating tissues recovered more quickly from blast and thermal effects than species relying upon seeds for reestablishment.

Some South Pacific test islands were altered by erosion or silting, and stands of coconut were eradicated or damaged by heat and blast and did not grow again. Radionuclide accumulation in the marine food chain was substantial, including cobalt 60. Terrestrial organisms accumulated soluble radionuclides, such as cesium 137 and strontium 90. Radionuclides in the South Pacific remained in the upper few centimeters of soil during the decade following testing. Many radioactive isotopes are as yet poorly studied for their genetic, cellular, and developmental effects, and for long-term effects on human reproduction.

In the Southwest, some annual plant species, such as *Salsola*, rapidly invaded some disturbed areas; in dominant shrubs, seedling establishment was limited by inhospitable environmental conditions and grazing by small mammals. Lizard and small mammal populations at the Nevada test range were adversely affected by long-term radiation exposure. Individual lifespans, time of sexual maturity, population age distributions, and other life-history characteristics were important determinants of population-level responses to ionizing radiation.

Radiation studies using either chronic or acute exposure regimes indicate that life-history stages or tissues undergoing rapid cellular division (e.g., embryonic and developmental stages) are most vulnerable to the adverse effects of ionizing radiation.[35] Comparing several categories of life forms, the following describes relative sensitivities (measured as acute lethal doses ranging from 200 for mammals to one million rads for bacteria): mammals are more sensitive than birds, which are more sensitive than higher plants, fishes, amphibians, reptiles, and crustaceans; these, in turn, are more sensitive than insects, bacteria, and viruses.[36] Of plant communities, coniferous forests, which dominate a large proportion of the Northern Hemisphere, are most sensitive; grasslands, characterized by plants with meristems protected near or below the soil surface, are least sensitive.[36]

In general, the long-term effects of ionizing radiation on natural ecosystems are difficult to assess. Studies using simulated fallout verify that surface deposition of radioactive particulates can alter the reproductive and growth potential of many plant species.[38] Each test disturbs the environment and "constitutes a selective biocidal agent that compromises the survival of some vulnerable species, while perhaps favoring the survival of others."[39] Similar studies focusing on decomposers have failed to show significant effects, except at very high exposures.[40]

Nagasaki and Hiroshima

Retrospective studies of Nagasaki and Hiroshima, while not weapons tests, offer new evidence of genetic damage in ferns exposed to radiation four and a half decades ago. The primitive reproductive systems of ferns are easy to examine, and much is known about fern genetics. American and Japanese botanists have reported that the offspring of the exposed plants show a variety of abnormalities and most do not mature successfully. Spores from ferns in the Nagasaki area produced mutants three times as often as those from ferns growing 25 miles away. Some spores tested were from survivors of the bombing in 1945, and others were from clones from those plants. The ferns may be an indicator of what we are about to learn.

The study was led by Edward J. Klekowski, Jr., from the University of Massachusetts, who speculated that undetectable genetic damage "was probably even worse" among other kinds of plants.[41] Since no prewar surveys of Nagasaki's plants exist, it is impossible to know whether the radiation altered the mix of species.

In both Nagasaki and Hiroshima, many trees and other plants lived through the atomic explosions, as others did later in the South Pacific. Specimens of numerous trees, including ginkgo, black locust, and camphor trees—some only 650 yards from the hypocenter of the blast—withstood the searing heat and are still growing. Immediately after the war, many Japanese observed some abnormalities on trees, but faced with monumental rebuilding tasks, they collected no systematic data on biota changes. Klekowski noted that apart from burn scars, most trees appeared healthy. But he said he "would bet on it" that some had sustained genetic damage and reproductive difficulties.[42]

UNDERGROUND TESTS

Protests over atmospheric testing drove testing underground. In 1963, following the Cuban missile crisis, Kennedy and Khrushchev quickly reached agreement on a surface test ban treaty, reflecting accomplishments of technical meetings initiated in 1958 by Eisenhower and Khrushchev. Tests were banned in the atmosphere, under water, and in outer space. The price Kennedy had to pay for support from the Joint Chiefs of Staff was continuation of tests underground.

The 1963 limited test ban did serve to contain surface damage and atmospheric injection of dust and radioactive particles, but it did not eliminate testing or prevent the further refinement of nuclear weapons. The kill efficiency or lethality costs are now estimated to be reduced to 12 cents for each death (one weapon for 10 million deaths). At the same time, the 1963 treaty made the costs of tests more expensive. One source estimated that costs climbed from a total of $24.6 million in 1980 to $726.9 million in 1987.[43]

Test costs are highly variable. The annual NTS budget is about $1 billion. About a total of 12 tests are conducted each year for each of two types of tests: weapons design and yield, and weapons effects on military equipment and hardware. New safety requirements mentioned below are increasing costs. Because of high costs, few tests are made for stockpile reliability. Tests are acknowledged to be essential for third-generation and SDI weapons.

Underground testing also has made verification and test ban agreements more complex and difficult to achieve, but advances in seismology and other verification techniques and recent official joint U.S.–U.S.S.R. tests indicate that underground weapons tests can be differentiated from earthquakes and verified down to acceptable margins of error. Resistance to further test ban agreements thus turns less on technical verification capability than on desire and political will.

Environmental Effects at the Nevada Test Site

Surveys[44] of the environmental effects of underground tests indicate more than just land subsidence, rock compression, fracturing, and test-induced earthquakes (of a lower magnitude than the weapons explosion). By 1977, there were 55 contaminated Radiation Exclusion Areas covering a total of 9 square miles, with surface radiation levels reaching at least 5 millirems per hour. Plutonium and other radionuclides have been found three feet deep in areas where the ground surface has been disturbed. Plutonium, windblown or animal transported, is also found at the surface in undisturbed and offsite areas with readings in the nanocurie (one billionth of a curie) to picocurie (one trillionth of a curie) per gram range. "Safety tests," begun in 1955 to assure that a weapons accident would not inadvertently happen, have contaminated areas with dispersed plutonium, uranium, and radioactive materials.

Until recently, it was thought that the short-term, direct effects (approximately 36 hours) of underground tests were relatively well known, and that only long-term effects (approximately 1 to 100 years) were poorly understood. NTS officials recently discovered they had erred: they overestimated rock strength. Tests have been conducted with unacceptably small margins of safety for years.

The Mighty Oak test of April 10, 1986 changed their view. Engineers tried to simulate the effect that a blast in space would have on an orbiting satellite or other space equipment. They exploded a device at one end of a long horizontal pipe and allowed the radiation to rush along the pipe toward the target equipment. The pipe was equipped with doors to shield the target from the blast's shock wave. The instant the radiation passed the doors, they were supposed to close.

In the test, the shock wave penetrated to the equipment; $32 million of equipment was destroyed; and an unreported, perhaps unknown, quantity of radiation was vented. Engineers believed at first that the doors had not closed. But in a January *New York Times*[45] interview, DOE's Nevada operations office in Las Vegas revealed it had found that the doors were closed. The shock wave instead was transmitted when the surrounding rock gave way. "We were much closer to the edge than we realized," said James K. Magruder, assistant manager for operations. Officials are said to have changed testing procedures to ensure that radiation does not escape. Stronger barriers and prefilling more of the tunnel pipe prior to the test are said to add more than $9 million per test.

A 1989 Office of Technology Assessment report[46] disclosed for the first time that before 1970, more than 12 billion curies of radiation had been released, and since 1970, 126 underground tests have released a total of roughly 54,000 curies of radiation. Some releases have been caused by design miscalculation, others by deliberate purgings of cavities containing reusable test equipment and by the drilling of holes near the explosion to obtain samples of molten rock. The nonpartisan group that participated in the OTA study stated that the radiation releases have not harmed the health of nearby residents. However, earlier Congressional hearings found that the radiological monitoring currently in place is inadequate to estimate properly the size and distribution of "offsite" releases.[47]

Water pollution is also a problem. Tritium, krypton 85, and sodium 22 above background levels have been detected in groundwater, and DOE recognizes the need for more wells and better information. DOE's budget for NTS environmental research is about $3 million to $4 million a year, and it is reported to cost about $1 million to drill a well, so not much is done. The costs are high because the water table is far below the surface. Groundwater, DOE estimates, could take 1,000 years to migrate offsite.

Senator John Glenn draws the critical conclusion. "Nuclear weapons have always been touted as giving us 'more bang for the buck,' " he notes. But the unpaid bill for environmental and health damage "demonstrates that they are far more expensive than we thought."[48]

CONCLUSION

The environmental effects of nuclear weapons testing are still best viewed as a ticking time bomb, and it is the height of irony that as relations between the United States and the Soviet Union show tangible signs of improvement, the Department of Energy still proposes to modernize and expand weapons production capacity and to conduct additional weapons testing. That it does so partly in the name of environmental compliance adds insult to injury. Their effort, however, is a clear indicator of a flawed approach to the larger problem of assuring national defense.

Weapons tests historically represent an effort to find a technical fix to the prime security problem—national vulnerability—by unilateral rather than by bilateral or multilateral means. It is a Sisyphean effort, however, for confidence in defense tends to be a function of the vulnerability problem. In a time of conflict, like the one we are just coming out of, a country deters an adversary politically and militarily by threatening harm and destruction, not simply by striving to shield itself against injury. Hence, the sword and shield metaphor. Vulnerability is thus double edged, for nations strive to offset their own weaknesses and to keep their adversaries susceptible to attack. Each nation attempts to outdo the other in its efforts, but technical advantage and unilateral fixes are invariably negated by counter-response efforts. It is vulnerability that provides the dynamics of the arms race.

Reagan's arms build-up in the 1980s offers an example of the futility of unilateral efforts to assure defense. Reagan may have made the United States militarily stronger with his budget breaking arms build-up, but in spite of his strengthening of American defenses the United States remains vulnerable today to a Soviet attack, perhaps even more so than at any time since the end of World War II. His ratcheting up of the American threat could not stop Soviet unilateral efforts to build up their weapons in numbers and in lethality, and to hold in jeopardy the United States ICBMs. The debates over SDI and the MX and Midgetman missiles are symptomatic of the flaw in an approach to the American vulnerability problem based on a unilateral and technical fix.

If so, a change of direction is essential and appears to be in the making. Under a new cooperative arrangement, the problems that drive the nuclear arms race can be dealt with: fear of a first-strike capability that negates confidence in a second-strike deterrent and that can in turn induce political paralysis; fear of new weapons that could alter the military balance; and fear of a clandestine stockpile of reliable weapons. In contrast with these fears is the need for confidence in survivable and reliable weapons to deter hostile political and military moves. What makes a system of better controls and lower thresholds of threat essential is that most arms control scenarios, including those of arms control organizations, do not foresee a nuclear-free world for several decades. Instead, most assessments appear to envision a future relationship built around a minimum deterrence system, with each nation possessing around 1,000 weapons to assure a second-strike system.

Two problems with traditional arms control approaches is that they are narrowly construed to deal with discrete problems rather than with the collective set of problems that drive the arms race, and they have also allowed the development of new weapons. For these reasons they have failed to prevent escalation in the numbers of warheads, to halt the development of new weapons, and to overcome uncertainties about weapons in the stockpile.

What is needed is a system that addresses all three of these problems in an interlocking, self-enforcing way. A system that can be called a Verification Triad (to play off of a deterrence triad) is boldly outlined here. Its key elements already exist in actual or proposed arms control arrangements and all can be realized with existing technologies. Its critical components include:

1. A START agreement that will bring about the reductions in delivery systems and warhead numbers necessary to overcome the first-strike fear, for the numbers of weapons are reduced to the level that does not threaten a second-strike deterrent.

2. A comprehensive test ban treaty that will block the development of new weapons, for new weapons have to be tested for reliability and effectiveness before they can be deployed. As noted above, recent negotiations and joint tests indicate both nations recognize that verification that discriminates between earthquakes and weapons tests down to acceptable levels of error is at hand.
3. Finally, a materials production ban plus verification controls on tritium, enriched uranium, or plutonium. These fuels are essential for nuclear weapons. A ban and control of weapons material production will reduce the amount of fresh fuel available for weapons and can have the twin effect of helping to overcome fear of a hidden stockpile of reliable weapons and of eroding a military planner's confidence in nuclear weapons.

Problems abound that relate to the kind of material used in weapons and to diversion and cheating. With tritium-dependent weapons, the half-life is short, and the rapid decay forces a reduction of weapons if no fresh tritium is produced. Weapons, however, can be built without tritium, and a sea of plutonium with a long half-life exists, so a productions ban alone will not be enough to overcome fear of a clandestine stockpile. What is essential is a credible verification system with on-site inspection provisions.

The 1988 Intermediate Nuclear Forces Treaty offers the anatomy of a model for verification controls to prevent diversion and cheating. Its key elements include: "baseline inspections" to determine and to continuously monitor the materials in the stockpile; "close-out inspections" to verify that prohibited activities and dedicated military reactors are shut down; "elimination inspections" to observe destruction of missiles, launchers, and warheads; "short-notice inspections" of suspect facilities; and "continuous portal monitoring" of weapons production and assembly plants to ensure no new weapons, triggers, or delivery vehicles are being constructed or shipped from assembly plants.

If interlinked, the control system becomes self-enforcing. It would help to build the confidence levels essential for a stable relationship, which is the sufficient condition for confidence in defense. This system offers a viable alternative to DOE's modernization plan; it provides a formidable system of controls but it does not require expanded capacity and continued weapons testing.

In addition, it should be matched with a set of measures to deal with other critical problems raised in this chapter. These measures should include:

- Environmental protection factored into decision cycles early, so that environmental need is not dealt with retroactively as a secondary concern.
- Nuclear weapons cost-effectiveness estimates that include environmental costs, not simply lethality. It is time to put to rest the notion that "nuclear weapons are cheap."
- A more accountable and responsive DOE created through stronger Congressional oversight and independent reviews of environmental impacts.
- U.S. and the U.S.S.R. discussions on environmental issues associated with weapons testing and materials production. Placing issues not now discussed on the table could help build confidence and reduce the threats of war, the costs of the arms race, and environmental harm.

All of these acts would help to assure greater accountability as well as ratchet down and constrain nuclear weapons development at a time when it appears that nuclear weapons will continue to exist for the forseeable future. Instead of pursuing efforts to secure technical and unilateral fixes to defense questions, which will probably prove as illusive as ever, we would be better served if we sought joint measures to reduce threats and to build to a climate of confidence in a stable relationship. A verification system designed somewhat around the elements proposed here could be a step in the direction of a better protected environment, and a safer and more secure world.

NOTES

1. Department of Energy "Complex 21," *Washington Post*, July 15, 1990, and Zbigniew Brzezinski, *Game Plan* (Atlantic Monthly Press, Boston, 1986), p. 165.
2. Herbert F. York, ed., *Arms Control: Readings From Scientific American* (W.H. Freeman & Co., San Francisco: 1973), p. 73.
3. Ibid., p. 75.
4. Ibid., p. 73.
5. Ibid., p. 74.

6. Quoted in Robert Del Tredici, *At Work in the Fields of the Bombs* (Harper & Row, New York: 1978), p. 179.

7. U.S. General Accounting Office, "Operation Crossroads Personnel Radiation Exposure Estimates Should Be Improved." The opening page of the document states its principal findings in a memo to Senator Alan Cranston. Quoted by Del Tredici, op. cit., p. 176.

8. S. Glasstone and P.J. Dolan, *The Effects of Nuclear Weapons*, 3rd ed. (U.S. Government Printing Office, Washington, D.C.: 1977).

9. Ibid.

10. S.T. Cohen, *The Neutron Bomb: Political, Technological, & Military Issues* (Institute for Foreign Policy Analysis, Washington D.C.: 1978).

11. A.B. Pittock, T.B. Ackerman, P.J. Crutzen, M.C. MacCracken, C.S. Shapiro, and R.P. Turco, *The Environmental Consequences of Nuclear War.* Volume I: Physical and Atmospheric Effects (John Wiley & Sons, Chichester, UK: 1985).

12. K.Y. Kondratyev, *Climate Shocks* (Wiley, NY: 1988).

13. Thomas Hamilton, *"Health Effects from the Nuclear Testing Program: A Review,"* (Paper presented at the International Scientific Symposium on a Nuclear Test Ban, Jan. 15-16, 1988, Las Vegas, Nevada).

14. Carl J. Johnson, "Cancer Incidence in an Area of Radioactive Fallout Downwind From the Nevada Test Site," *Journal of the American Medical Association (JAMA)*, Jan. 15, vol. 251, no. 2, 1984.

15. Ibid. p. 230.

16. Ibid.

17. Ibid.

18. Lawrence Spohn, "Los Alamos Scientists Explore the Effects of the Bomb: Little Boy Replica Gives New View on Today's Radiation Hazards," *Albuquerque Tribune*, Friday, May 2, 1988, p. D-1.

19. Harvey Waterman and Norman Solomon, *Killing Our Own*, (Delta, New York, 1982).

20. York, op. cit., pp. 46-47.

21. Johnson, op. cit., p. 231.

22. Quoted in *Greenpeace* July/August 1988, p.7.

23. Joseph Bauman, "Downwinders: America's Nuclear Sacrificial Lambs," *High Country News*, July 6, 1987, vol. 19, no. 12, p. 10.

24. Ibid.

25. Ibid.

26. Ibid.

27. Ibid.

28. Ibid.

29. Johnson, op. cit., p. 231.

30. Bauman, p. 10.

31. Ibid. p. 16.

32. Ibid.

33. Ibid.

34. Herbert D. Grover and Mark A. Harwell, "Biological Effects of Nuclear War II: Impact on Humans." *BioScience,* October 1985, vol. 35, no. 9, p. 572.

35. G.M. Woodwell, "The Biotic Effects of Ionizing Radiation." *Ambio,* 1982 11: 143–148.

36. F.W. Whicker and V. Schultz, *Radioecology: Nuclear Energy and the Environment* (CRC PRESS. Boca Raton, FL.: 1982), vols. 1 and 2.

37. Ibid.

38. D.J. Nelson, ed., *Radionuclides in Ecosystems, Proceedings of the Third International Symposium on Radioecology. Conf-7100501,* 1971.

39. Grover and Harwell, op. cit.

40. D.J. Nelson, ed., op. cit., pp. 1222–28, cited in Grover and Harwell.

41. Erik Eckholm, "Genetic Harm to Plants Found After Atom Blast," *New York Times,* Friday, Aug. 9, 1985, p. 4.

42. Ibid.

43. Natural Resources Defense Council, "*New NRDC Study Reveals 71 Unannounced U.S. Nuclear Weapon Tests,*" *News Release,* Jan. 16, 1988 (Washington, D.C.).

44. Dana Coyle et al., *Deadly Defense: Military Radioactive Landfills.* A Citizen Guide by Radioactive Waste Campaign (New York: 1988).

45. Matthew Wald, "Errors Cut Safety Margin in Bomb Tests," *New York Times,* Feb. 1, 1989, p. A-11.

46. R. Jeffrey Smith, "Underground N-Tests Basically Harmless, Study Says," *Albuquerque Journal,* Nov. 3, 1989, p. B-4.

47. Michael McCally and Christine K. Cassel, *The Nuclear Weapons Test Ban in 1988,* A Report Presented to the International Scientific Symposium, Las Vegas, Nevada, Jan. 14, 1988 (5624 S. Dorchester, #3, Chicago, Ill., 60637), March 10, 1988.
48. Matthew Wald, "*Bomb Site Cleanup Is Put at Billions.*" *New York Times* (national edition), July 2, 1988, p. 7.

7

If Deterrence Fails: Nuclear Winter and Ultraviolet Spring

JOHN W. BIRKS
ANNE H. EHRLICH

PREVIOUS CHAPTERS IN THIS BOOK HAVE DEALT WITH THE SEVERE AND GROWING environmental consequences of developing, deploying, and maintaining weapons of war, especially nuclear weapons. The sole justification given for the accumulation of these weapons is the prevention of war. In theory, the approximately 50,000 nuclear warheads in the arsenals of the NATO and Warsaw Pact countries will never be used; the threat of mutual annihilation presumably makes nuclear war impossible. Our history of more than forty years of deterrence is often cited as evidence that nuclear weapons have served this function, but it is hardly reassuring when one considers that several hundred local wars have been fought during this short period of time. In fact, the use of nuclear weapons has been only narrowly averted on numerous occasions, and four decades is a very short time in comparison with either the span of recorded history (~5,000 years), or what we hope to be the future of our civilization (reader's estimate here).

The principal goal of this book is to call attention to the severe damage inflicted on our planet every day as a result of maintaining the nuclear standoff known as deterrence. In this chapter, however, we take a look at the catastrophic damage that would result if nuclear weapons actually were to be used in substantial numbers. The immediate human deaths and injuries, as well as damage to buildings and infrastructure, caused by the direct effects of nuclear explosions—the thermal pulse, blast, and radioactivity—have been clear since Hiroshima. The probability of atmospheric effects and serious effects on ecosystems and agriculture also has been known.[1] But the devastating effects on the Earth's atmosphere and global climate have only recently been quantified and brought to public attention.[2] It is now recognized that the long-term effects of a nuclear war on agriculture and ecosystems are likely to result in as many or more human deaths than the prompt effects of the nuclear fireballs.

In this chapter we will begin by describing our current understanding of the physical changes in the environment that would result from a major nuclear war. This will be followed by a description of the likely detrimental effects of these changes on agriculture and ecosystems.

NUCLEAR WINTER

There is now a consensus that a major nuclear war would be followed by a period of cold and darkness popularly known as "nuclear winter." Although uncertainties still remain, the qualitative conclusions of the early work on this problem[3] have been confirmed by studies of the U.S. National Research Council,[4] the Scientific Committee on Problems in the Environment (SCOPE, a committee of the International Council of Scientific Unions)[5] and the United Nations.[6]

Described simply, nuclear winter is a result of smoke injected into the atmosphere by the enormous fires ignited by nuclear explosions. Smoke particles strongly absorb visible radiation from the sun and thereby prevent the warming rays of the sun from reaching the Earth's surface. The result is a heating of the upper atmosphere by as much as 100°C (180°F) and an average cooling of a large fraction of Northern Hemisphere land surfaces by more than 10°C (18°F) for a nuclear war that occurs in the summer. Such a temperature reduction would result in localized outbreaks of subfreezing temperatures over large land areas.

Fires and Smoke

The severity of nuclear winter depends mostly on the amount of smoke produced. More specifically, it depends on the amount of "soot" or elemental carbon introduced to the atmosphere by the fires. Sooty smoke, because it is black, is highly absorbing of visible light and is produced in fires involving petroleum and petroleum products such as asphalt roofing materials and plastics. The world inventories of these materials have been estimated to be 800 to 1,500 teragrams (1 Tg $= 1 \times 10^{12}$ grams $=$ 1 million metric tons) of petroleum and related products, 200 to 500 Tg of asphalt roofing, and 340 to 460 Tg of plastics.[7] Typically, about 10% of the weight of these materials is not fully combusted, but enters the atmosphere as very black smoke particles. Construction lumber has a smaller "soot emission factor" (about 1-3%),[8] but there is much more of it available for burning. Estimates of the global inventory of lumber and related products range from 6,000 to 13,000 Tg.[9] Forests and grasslands would be impacted by nuclear explosions, but the overall contribution to the total smoke produced has been estimated to be quite small.[10]

Scenarios

Because nuclear winter depends on the amount of smoke that enters the atmosphere, the predicted severity of nuclear winter is highly scenario dependent. The nuclear warheads may be considered to be the "matches" that ignite cities and oil refineries. Thus, the number of megatons of explosive yield used is not critical; the nature of the targets against which the weapons are directed is much more important in determining the amount of smoke produced. For example, it has been pointed out that 100 megatons (Mt), which is less than 1% of the approximately 13,000 Mt available in the arsenals of the NATO and Warsaw Pact countries, would be sufficient to produce severe climatic effects.[11] This requires that the 100 Mt be composed of 1,000 individual 100-kiloton (about eight times the size of the Hiroshima bomb) warheads and that these warheads be used against the 1,000 largest cities in the Northern Hemisphere (approximately all cities of 100,000 population or more). Nuclear winter could be produced by an even smaller number of warheads if they were directed only against oil refineries; again, the amount of sooty smoke produced is what counts.

Of course, it is extremely unlikely that the first 100 Mt of a nuclear arsenal would be used against cities or oil refineries; instead, they almost certainly would be directed against military targets such as missile silos, air and submarine bases, and command and communication centers. Military targets would consume at least the first 1,000 Mt of warheads from each side.[12] These targets, especially missile silos, have much smaller fuel loadings, and the argument has frequently been made that a "limited nuclear war" in which only military targets are attacked would not have significant climatic effects. The point often debated here is whether or not a limited nuclear war can remain limited. If the war were escalated to include population centers—as it probably would be, in our opinion—then a nuclear winter would be inevitable.

In order to defuse the often emotional debates surrounding scenarios, it is useful to consider a range of possibilities and to cast these scenarios in terms of amounts of smoke injected rather than in terms of numbers of warheads, total megatonnage, target distribution, and the like. At its 1987 Bangkok workshop, the SCOPE group devised three scenarios for smoke injection during a nuclear war, and these have now been analyzed by several climate modeling groups. The scenarios provide for the injection of 15 Tg, 50 Tg, and 150 Tg of smoke into the atmosphere, respectively. Furthermore, since the altitude of injection is important to the model calculations, the height distributions of the injections are varied also. The centers of mass of the smoke injections are 2.8 kilometers (km), 4.0 km, and 6.7 km above sea level, respectively, for the 15 Tg, 50 Tg, and 150 Tg scenarios. This smoke is assumed to be approximately 50% soot (elemental carbon) with an absorption coefficient at visible wavelengths of 5 m^2/g (meters squared per gram). The latter is a quantitative measure of the degree to which the particles will absorb light.

These scenarios are believed to span the range of uncertainty in the amount and distribution of soot that would enter the atmosphere as a result of a nuclear war. The 50 Tg case is similar to that assumed in the 1985 NRC study,[13] which estimated a total of 180 Tg of smoke, but this smoke was considered to be only 20% soot. A recent detailed analysis by Small, Bush, and Dore found that 40 Tg of smoke would be produced in a full-scale attack on the U.S. alone.[14] A recent, independent estimate by Soviet scientists[15] predicts 70 to 90 Tg of smoke based on the 1985 NRC target distribution. Thus, the best estimate at present is that the climatic effects of a major nuclear war would lie between the 50 Tg and the 150 Tg Bangkok scenarios. A more limited nuclear war might correspond to the 15 Tg case.

Darkness

The soot particles injected into the atmosphere by the thousands of fires ignited by nuclear explosions in and near cities, oil refineries, and forests would be sufficient to block most of the sunlight from reaching the

The approximate attenuation of sunlight reaching the Earth's surface by the overlying smoke layer is easy to calculate according to the Beer-Lambert Law using a hand calculator. The transmission, T, of sunlight through the smoky layer is given by

$$\frac{I}{I_0} = e^{-m\sigma \sec(\phi)/A}$$

where m is the total mass (in grams) of smoke injected into the atmosphere, σ is the absorption coefficient of the smoke(5 m^2/g), A is the total area over which the smoke is distributed (in square meters), and ϕ is the solar zenith angle, i.e., the angle between the sun and a line perpendicular to the Earth's surface. At 45° north latitude, the average value of the solar zenith angle is 54.5° in midsummer.

An analysis of the likely targets of a nuclear war indicates that most of the nuclear-war fires will occur in the middle half of the northern hemisphere (~ 30-70° north latitude, 120 million km^2). East-West dispersion of smoke clouds is expected to occur on a time scale of a few days, as is true of the normal atmosphere, while North-South (latitudinal) dispersion occurs much more slowly. As an illustrative calculation, we will assume that the smoke becomes uniformly distributed over half the Northern Hemisphere. The results (which the reader can verify) are:

Scenario	Transmission, T
15 Tg	0.34
50 Tg	0.027
150 Tg	0.000021

Box 1: Sunlight Obscuration

Earth's surface over large areas of the globe. Calculations of the average transmission of sunlight through the atmosphere for the three Bangkok scenarios are described in Box 1. These calculations indicate that on the average, only 34%, 3%, and 0.002% of the solar radiation would reach the Earth's surface for the 15 Tg, 50 Tg, and 150 Tg scenarios, respectively. Note that even for the limiting 15 Tg scenario, the solar intensity would be reduced by more than 50%.

Such calculations do not take into account light scattering by the particles, which both reflects some light back to space and increases the path length of other rays through the atmosphere so that they are more likely to be absorbed. When complete multiple scattering calculations are carried out using computers, the average amount of sunlight reaching the surface is reduced even further. Of course, the smoke will always be somewhat patchy, so periods of less and greater darkness than the averages estimated in Box 1 would occur.

Surface Cooling

It is not immediately obvious that the introduction of soot to the atmosphere would cause the Earth's surface to cool. In fact, because the albedo (reflectance) of the Earth would be reduced (much of the Northern Hemisphere would appear dark from space) the Earth's atmosphere system would in fact trap more incoming solar radiation and heat up. In order to understand why the surface actually cools, one must consider the "greenhouse effect."

Without the greenhouse effect, our Earth would be a frozen planet. If the Earth had no atmosphere, the average surface temperature would be about −18°C (0°F), well below the freezing point of water, which is 0°C or 32°F. This equilibrium temperature results when a balance is struck between the absorption of visible light from the sun and the emission of infrared light by the Earth. The average temperature of the Earth is actually much warmer, about 15°C (59°F), because much of the infrared radiation emitted by the Earth's surface is reabsorbed in the atmosphere, principally by small amounts of water vapor and carbon dioxide (CO_2). A current, long-term environmental concern is that the buildup of CO_2 in the atmosphere from the burning of fossil fuels will eventually cause a major warming of the Earth and change our climate.

Nuclear winter would have just the opposite effect. Because most of the sun's visible light would be absorbed high in the atmosphere, the natural greenhouse effect would be "turned off" and the temperature of the Earth's surface would decline rapidly. Fundamental to this negation of the greenhouse effect is the circumstance that although soot absorbs very strongly at visible wavelengths, it is a very weak absorber of infrared radiation. Thus, soot clouds would block the warming, visible radiation from the sun from reaching the surface, while allowing infrared radiation emitted by the Earth to pass through. As a result of this imbalance, the surface of the Earth would cool.

Actually, in a nuclear winter only the land surfaces of the Earth would cool substantially. This is because the upper layer of the oceans, which is about 100 meters deep, is well mixed and therefore has a very large heat

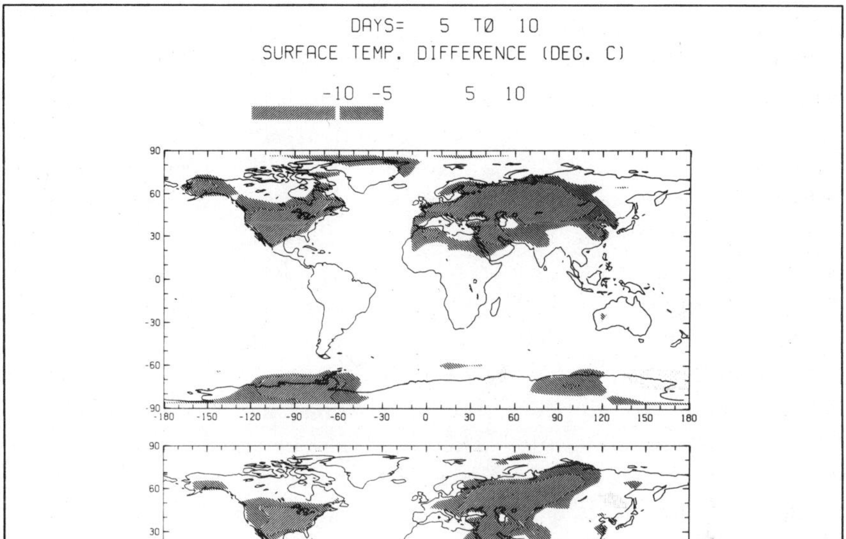

Figure 7-1: Surface temperature changes for the three Bangkok nuclear war scenarios as calculated in a General Circulation Model by G. Glatzmaier and R. Malone of Los Alamos National Laboratory. The three scenarios are 15 Tg soot (top), 50 Tg soot (middle), 150 Tg soot (bottom) injected. The temperature maps correspond to the difference between the average temperature of days 5 through 10 of the simulation and the normal monthly mean temperature. Note that the warm and cool regions near Antarctica are manifestations of storms that occur naturally in this region. These effects would average out over a longer period. Figure courtesy of R. Malone and G. Glatzmaier.

capacity. Since the land surface is not fluid, a loss of heat on only the top few centimeters causes the temperature to decline. To change the temperature of the oceans, it is necessary to cool the entire mixed layer. Smoke is not expected to persist in the atmosphere for a long enough period of time for the oceans to cool substantially. Average sea temperature reductions are projected to be less than 2°C, while local temperatures could decline by as much as 5°C for a few months.[16] The warm oceans tend to have a buffering effect on land temperatures as well. Air parcels moving from ocean to land would tend to warm the adjacent land surface, and cooling is therefore predicted to be greatest in continental interiors.

Figure 1 shows the reductions in temperatures for days 5 through 10 following a July nuclear war for the three Bangkok scenarios as calculated by the general circulation model (GCM) of Glatzmaier and Malone at Los Alamos National Laboratory.[17] This GCM has been made fully interactive with the smoke particles so that their heating of the atmosphere actually changes the wind fields, cloud cover, precipitation, and other conditions. Average surface temperature reductions for this five-day period over land between 30 and 70 degrees north latitude are 5°C (9°F) for the low smoke case, 14°C (23°F) for the medium case, and 22°C (40°F) for the high case. Maximal drops in temperature within continental interiors exceed 30°C (54°F), and local freezing events are widespread during the first month following the nuclear war for both the intermediate and high smoke cases.

Climate models predict a much smaller cooling for a nuclear war that occurs in the winter. The reason for this is that in winter the intensity of sunlight falling on land sufaces is already considerably reduced. Thus, the blockage of sunlight by a smoke layer decreases the solar intensity in winter by a smaller amount than in summer, and the effect on temperature is less. Said another way: in winter it is already cold. Furthermore, any temperature reductions in wintertime will have a much smaller impact on the biosphere than temperature reductions that occur in summer. For obvious reasons, a war in spring or fall would produce smaller temperature drops than in summer but larger than in winter. The significance for living systems would be greater in spring than fall and would depend in part on how long the effect persisted. Of course, one cannot predict the season in which a nuclear war might occur.

Drought

One general prediction made by all 3-D climate models is a large reduction in precipitation following a nuclear war. This is a result primarily of the stabilization of the atmosphere (upper troposphere) by solar heating of the smoke layers. In the natural atmosphere, parcels of moist air are warmed at the Earth's surface and rise to higher altitudes, where it is colder. The result is condensation of water vapor to form clouds and precipitation. In the nuclear-winter atmosphere, hot smoke layers overlie cold air at the Earth's surface. This so-called "temperature inversion" is very stable against vertical mixing, and as a result precipitation is reduced. In the GCM of the Lawrence Livermore National Laboratory, the decrease in rainfall over land for a July nuclear war is calculated to be about 75% (averaged over days 20 to 30 of the simulations) for all three Bangkok scenarios.[18] A very significant forecast in most GCMs is a failure of the summer Asian monsoon.[19]

ULTRAVIOLET SPRING

The millions of tons of soot and toxic chemicals injected into the atmosphere would change not only the energy balance of the Earth/atmosphere system but also its chemistry. Some of the chemical effects to be expected include acid precipitation, photochemical smog formation, widespread distribution of chemical toxicants, and depletion of the Earth's protective ozone layer. Of these, probably the most damaging to plants and animals would be the reduction in the ozone shield.

Stratospheric Ozone

The stratosphere is a very stable layer of the atmosphere at an altitude of about 12 to 50 km (7 to 30 miles). The stratosphere contains very low concentrations of the molecule ozone, O_3, which serves the important function of absorbing biologically damaging ultraviolet radiation (UV-B) from the sun and thereby shielding the Earth from the damaging effects of UV-B on organisms. The absorption of ultraviolet light by ozone heats the stratosphere and creates a temperature inversion (temperature increasing with increasing altitude), which defines this layer of the atmosphere and accounts for its extreme stability with respect to vertical mixing.

Ozone is constantly being formed high in the midlatitude stratosphere by the photolysis (splitting caused by sunlight) of oxygen molecules (O_2) to form single oxygen atoms. These oxygen atoms attach to other oxygen molecules to form ozone (O_3):

$$O_2 + h\nu \rightarrow O + O$$

$$O + O_2 + M \rightarrow O_3 + M$$

Here, $h\nu$ (Planck's constant h times the frequency of light ν, the energy of a photon) indicates the absorption of a photon of light. For oxygen this can occur at wavelengths shorter than 242 nanometers. The M in the second reaction is any molecule, usually N_2 or O_2, and is necessary to carry away the excess energy when the new chemical bond is formed.

If these reactions were the only reactions occurring in the stratosphere, all of the oxygen in the atmosphere would soon turn into ozone. In fact, there are numerous reactions that destroy ozone. One of the most important is the sequence of reactions:

$$NO + O_3 \rightarrow NO_2 + O_2$$

$$NO_2 + O \rightarrow NO + O_2$$

$$O_3 + h\nu \rightarrow O_2 + O$$

$$\text{Net: } 2\,O_3 \rightarrow 3\,O_2$$

These reactions constitute a catalytic cycle. Each nitric oxide molecule consumed in the first reaction is regenerated in the second reaction; as a result, one nitric oxide molecule can cause the destruction of large numbers of ozone molecules during its residence time in the atmosphere. It is now well established that this sequence of reactions controls the amount of ozone in the tropical and mid-latitude stratosphere. The NO naturally present in the stratosphere is derived from nitrous oxide, N_2O ("laughing gas"), which is produced by sea and soil bacteria in both nitrification and denitrification processes. Thus, bacteria at the Earth's surface help control the amount of ozone in the stratosphere.

As described in the text, nuclear explosions heat air to very high temperatures and thereby form nitric oxide, NO. Some of the nitric oxide would be injected directly into the stratosphere by the rising fireball. However, since most nuclear explosions would have a yield of less than 0.5 Mt, a large fraction of the nitric oxide initially would be deposited beneath the stratosphere in the altitude region 5-12 km. Much of this NO would then be lofted into the stratosphere by solar heating of the soot clouds. Once in the stratosphere, nitric oxide would add to the rate of ozone destruction via the catalytic cycle described above.

Box 2: Ozone Photochemistry

The ozone layer is quite fragile and susceptible to destruction by chemical pollutants. For example, the recently discovered Antarctic ozone hole is believed to be a result of the release to the atmosphere of chlorofluorocarbons (especially $CFCl_3$ and CF_2Cl_2) used as refrigerants, aerosol propellants, foam blowing agents, and industrial solvents. Part of the reason that the ozone layer is so fragile is that there isn't much of it. At its peak concentration, ozone is only about 10 parts per million of the total air molecules present, and if one were to compress the ozone in the stratosphere to a shell at atmospheric pressure it would be only about 3 millimeters (one-eighth of an inch) thick.

Some of the most important chemical and photochemical reactions that regulate the quantity of ozone in the stratosphere are described in Box 2. Ozone in the stratosphere exists in a dynamic steady state. It is constantly being formed by the interaction of ultraviolet light from the sun with oxygen, and constantly being destroyed by reactions with ultratrace quantities of various reactive chemical species, particularly the oxides of nitrogen. The chemical species nitric oxide (NO) is the most important catalyst for ozone destruction. In fact, the presence of low part-per-billion levels of NO limits the amount of ozone in the stratosphere to about half of what it would be if no NO were present.

Nitric Oxide from Nuclear Explosions

Nuclear explosions heat air to very high temperatures—of the order of 20 million°C. At these temperatures all molecules are completely dissociated into atoms and ionized. As the fireball cools by radiative emission and entrainment of additional air to form the "mushroom" cloud, the atoms of nitrogen and oxygen recombine; however, not only are the air molecules N_2 and O_2 reformed, but also the molecule NO (nitric oxide) is produced. Of all air molecules that are heated to at least 2000°C, about 1% will be reformed as NO. The hot fireball rises in the atmosphere, carrying this nitric oxide with it. If the yield of the nuclear explosion is 1 Mt or larger, the fireball will completely penetrate the stratosphere.

It has been recognized since 1973 that nuclear explosions would increase the amount of nitric oxide in the stratosphere by a large factor and cause an erosion of the ozone shield.[20] This concern prompted a 1975 study by the U.S. National Research Council.[21] The discovery of the nuclear winter effect, however, has further exacerbated the problem. GCM calculations, since they have been made fully interactive with smoke particles, now show that the solar-heated smoke layers will rise into the stratosphere. The smoke carries with it nitric oxide formed in nuclear explosions of yields too low to reach the stratosphere. Perhaps more important, the rising smoke clouds force ozone in the stratosphere to higher altitudes and across the equator into the Southern Hemisphere. The presence of soot also causes the stratosphere to heat by temperatures of up to 100°C (180°F), which in turn speeds up chemical reactions that destroy ozone, such as those involving the oxides of nitrogen.

The result is a very large "nuclear winter ozone hole" in the Northern Hemisphere, which is expected to persist for several years. The complete stratospheric chemistry has not yet been introduced to GCMs, but preliminary estimates indicate an ozone depletion in excess of 50%.[22] Such an ozone hole would result in a very large increase in the amount of UV-B radiation reaching the Earth's surface: roughly double to quadruple the normal amount. Of course, as long as large quantities of soot remain in the atmosphere, this radiation would be attenuated in the same way as visible light. As the soot is slowly removed from the atmosphere by rainout, however, sunlight would once again begin to penetrate to the Earth's surface. But this would not be ordinary sunlight: it would be highly enriched in UV-B radiation. Thus, a nuclear winter would be followed by an "ultraviolet spring."

CHRONIC PHYSICAL EFFECTS

The finding that a large fraction of the nuclear winter smoke would rise into the stratosphere, where its residence time is of the order of a year or more, greatly extends the time period over which climatic effects are to be expected. GCM calculations show that after only a few weeks, the smoke would spread over the entire Northern Hemisphere and deep into the Southern Hemisphere as well. As a result, significant, long-term climate perturbations should be expected. For example, climate modeling results at the National Center for Atmospheric Research indicate that cooling persists at the level of a few degrees for two years following the onset of a nuclear war.[23] Further investigation of these chronic effects will require that GCMs be made interactive with oceans and ice cover. At present, no conclusions can be drawn about these effects and how long they would persist.

CONSEQUENCES FOR BIOLOGICAL SYSTEMS

Nuclear winter phenomena following a large-scale nuclear war would have severe consequences for both agricultural systems and the natural ecosystems that underpin them. These longer-term problems would be above and beyond the devastation caused directly to any natural or agricultural areas near target zones by blast, heat, and local radiation.

Worse, the atmospheric effects of such a war would respect no boundaries; noncombatant nations, especially those in the most-affected 30 to 70 degrees north latitudes, would also be exposed to the atmospheric effects. It is seldom appreciated that nearly half the world's population—more than 2 billion people—live in those northern mid-latitudes, which encompass nearly all of North America, Europe, the Soviet Union, Japan, and much of the world's most populous nation, China. Still less appreciated is that the bulk of the world's food—over 60 percent—is grown there. Millions of people in tropical and subtropical nations depend on the productivity of the north for some portion of their food.

Thus, even without climatic consequences, a large-scale war in the Northern Hemisphere would cut off much of the world from vital food supplies. And, although the worst climatic consequences of a nuclear war would be visited on the world's breadbasket, they are expected to spread as well over the northern tropics and subtropics, affecting the next most populous latitudinal belt, with almost 2 billion people.

Even the much smaller populations of the Southern Hemisphere would not go untouched, although the reversal of seasons might be a mitigating factor for food production. Southern temperate-zone nations currently export food, but the amount of food they can produce even under ideal conditions could not begin to make up for the deficits caused by a loss of the Northern Hemisphere's harvest, even if food transfers were still possible.

Direct Effects

The immediate consequences to target populations of a nuclear war have been fairly well known since the first bombs fell on Hiroshima and Nagasaki. In a large-scale thermonuclear exchange, hundreds of millions of people could be killed or fatally injured, depending on the scale of the war and the targeting pattern. Human survivors in combatant nations would have to cope with their own injuries and radiation sickness, with no hope of help from the outside, and with much of their habitation, infrastructure, and transport systems in ruins. Moreover, serious damage would be inflicted on local biological systems that supply food and other essentials, and locally stored food would be destroyed or rendered unusably radioactive.[24]

Nuclear Winter Effects on Flora and Fauna

The persisting and spreading nuclear clouds would compound the already grave difficulties of recovering from such a horrendous event in target areas, while carrying the climatic effects to distant areas. The elements of a nuclear winter are so complex that it is useful first to examine their separate effects on flora, fauna, and ecosystems—so far as they are known or deduced—before attempting to assess total impacts. Many of the effects (such as low temperatures and low light levels) are known to interact synergistically in their impacts on biota, most often worsening the total impact. Additional unsuspected synergisms are entirely possible.

Reduced Light Levels and Temperatures. Green plants are the foundation of life for virtually all animals, including human beings. Through the process of photosynthesis, green plants and some micro-organisms (mainly bacteria and algae) capture energy from sunlight and convert it into the energy of chemical bonds in their tissues. Energy thus stored can be used by the "producers" themselves and becomes available to other organisms that feed on them (or that eat other consumers). Without sunlight, producers cannot photosynthesize and therefore cannot grow or reproduce. If totally deprived of light for long, they die. At light levels of roughly 10 to 20 percent of normal, most higher plants (including most crops) can maintain life, but cannot grow or produce seeds.[25] If light is reduced to only a small percentage of normal levels, many annual plants die within a few weeks.

Low temperatures also damage plant life. Many annual plants, including most crops, would be severely injured or killed by temperatures near or below freezing (0°C). Trees that normally withstand extreme subzero temperatures in winter can be seriously damaged by even a brief freezing episode during the growing season. Even a small temperature decline slows the rate of photosynthesis, which in a nuclear winter would add to reductions caused by low light levels.[26]

Temperate-zone plants have evolved ways to survive cold temperatures and low light levels in winter. Biennial and perennial plants spend winter months in a dormant stage; annuals die, leaving seeds to germinate in spring. Before the cold sets in, however, perennial plants go through a "cold-hardening" process—often triggered by changing day-length—in which they physiologically adjust to winter conditions over a period of several weeks.

Nuclear winter, by contrast, would produce low temperatures in a very few days, far too quickly for most plants to adjust adequately. They would try, however, and the rapid adjustment would be stressful, consuming significant amounts of energy. As patchy nuclear clouds passed over, light levels and temperatures would fluctuate between extremely dark and cold to nearly normal conditions. Plants would alternately attempt to cold-adapt and restore normal photosynthesis, consuming energy in each stressful change. Sooner or later, a plant's limited energy stores would be exhausted, and it would die. The patchy clouds and rapidly shifting conditions now predicted for a nuclear winter could be nearly as damaging to plant life as the persistent, uniformly far-below-freezing temperatures described in the first simple models.

Plants in subtropical and tropical regions are even more vulnerable to prolonged reduced light levels and low temperatures than are those in temperate areas, because they have no individual or evolutionary experience with such conditions. Even a sudden chill for a few hours can injure or kill sensitive tropical plants, which usually are unable to cold-harden significantly.[27]

Even more susceptible to low light levels than terrestrial plants are marine photosynthesizers, mainly algae. Such tiny organisms have almost no capacity for energy storage and would perish after a few days without sunlight. While the oceans' thermal inertia would prevent a rapid or extreme drop in water temperature, the loss of light would be enough to devastate algae and the marine fauna that depend on them, directly or indirectly.

Like plants, temperate-zone land animals have physiological mechanisms that enable them to survive a normal winter without undue stress. These devices include extended periods of hibernation or a less extreme lowering of metabolism and activity, withdrawal to underground areas, and inactive life stages (characteristic of many insects). Many birds and some bats and insect species migrate to areas of warmer weather. Most mammals grow fat layers and thick fur, and nonmigratory birds grow feathers, as insulation against winter cold. As with plants, the winter adaptations of animals take time to establish and are usually entrained by seasonal cues such as changing day-length. Mammals also must be able to build up stored energy (usually as fat) during summer months in order to survive the food-scarce winter.

The cold weather of nuclear winter, if occurring in spring or summer, would have severe effects on unprepared animal populations, which would have little capacity to cold-adjust or survive for long without food. Herbivorous animals, particularly those specializing on susceptible food plants, would soon find their food supplies dwindling as many plants died from the combination of chilling and too little light. Animals dependent on seeds or fruits would also face starvation as surviving plants failed to produce seeds. Carnivores might initially have an abundance of cold-stored food but would face starvation later. Winter conditions out of season would kill many insects caught in the wrong life-stage. And, like tropical plants, tropical animals have very little resistance to cold weather.

Reduced Precipitation. Marked changes in precipitation can have severe consequences for plant life, as anyone who has experienced a serious drought or a hurricane can testify. Plants generally are closely adapted to rainfall patterns where they grow; crops are, if anything, even more sensitive to too much or too little rainfall during the growing season. A midsummer reduction in precipitation by 75 percent, as is indicated in nuclear winter simulations, would be extremely disruptive of food production, although (unlike the searing drought of 1988) the lowered temperatures could have a somewhat mitigating effect.

Lack of water, however, could lead to the demise of many animals as streams, ponds, and other usually dependable water sources dried up. Livestock would be vulnerable in situations where their human managers were unable to provide water.

Air Pollutants. The explosions and fires of a nuclear war would release into the lower atmosphere a variety of noxious compounds, including halogens, furans, dioxins, and the usual components of air pollution such as oxides of nitrogen and sulfur—and in greater than normal quantities. Extremely acid precipitation is one likely result, adding the insult of acidification of ecosystems and croplands to the already severe injuries of chilling and darkness.

Ionizing Radiation. Radioactive fallout in the short and medium term would have serious consequences for many organisms, especially any in unsheltered areas near and downwind of target zones. Human beings happen to be among the organisms most highly sensitive to radiation, along with other "warm-blooded" animals. This sensitivity obviously includes livestock and poultry. Some plants, including coniferous trees and certain crops such as wheat, are also quite susceptible to radiation damage. High doses of radioactivity can kill. Sublethal doses can inhibit photosynthesis and cause genetic damage in plants. In animals, immune systems are suppressed, leading to greater susceptibility to diseases; genetic damage could produce an increased incidence of cancers and birth defects later.[28]

Radiation-damaged forests, grasslands, and croplands would also present a longer-term danger. The plants' resistance to damage from other nuclear winter effects—not to mention attack from pests and diseases—would be impaired, and many would be killed. Extensive tracts of land covered by dead or dying vegetation, especially forests, would be vulnerable to wildfires, and the fires would inject still more smoke into the atmosphere.

In natural ecosystems that escaped burning, differential susceptibility to radiation among organisms (which can be several orders of magnitude) would lead to highly skewed survival rates, even without accounting for different degrees of sheltering. For instance, animals living underground are more protected than those living in the open. Even a modest shelter, such as an ordinary building, affords a significant amount of protection against radioactivity, reducing the dosage by a factor of two or three. A substantial shelter, such as a well-sealed basement or concrete shelter, can have a protection factor of several hundred or over a thousand.[29]

Effects of Enhanced UV-B. Ultraviolet radiation is known to be potentially capable of causing damage across the entire spectrum of life; it can disrupt the most basic material of life, DNA, the genetic material, which governs development and reproduction in all life forms. UV-B thus can be both a powerful mutagen and a carcinogen. That it causes skin cancers, including some lethal ones, in human beings is well known. Unfortunately, less is known about details of other effects on human or other forms of life, although UV-B (like radioactivity) does depress rates of photosynthesis, as well as having other effects on growth and development in plants. In animals, UV-B acts to inhibit immune reactions.[30] Further research may well reveal a variety of harmful effects on many different organisms.

Combined Effects of Nuclear Winter Phenomena

Each nuclear winter effect by itself—sharply reduced and unpredictably fluctuating light levels and temperatures, altered precipitation patterns, air pollutants, acid deposition, radioactive fallout, and increased UV-B—

would have severe consequences on many forms of life and, both directly and through differential effects on species, for ecosystems. Combined, the results would be devastating. Damage to plants from cold and lack of light would be exacerbated by drought and exposure to air pollutants, acidity, radioactivity, and UV-B radiation, each of which can cause direct damage to plant tissues and/or inhibit photosynthesis. Animals that survived the cold would face starvation as well as assaults from radioactivity, UV-B, and air pollutants. In severe drought or freezing conditions, lack of water could be an additional problem.

The ability of ecosystems to function and provide life-support services for civilization would be heavily compromised by such conditions. Populations of many vulnerable organisms would certainly perish, with repercussions throughout food webs.[31] Herbivorous food specialists would disappear in the wake of their food-plants; their predators would also soon find slim pickings. Bees and other insects that are essential for pollination of numerous plants (including dozens of crops) would be vulnerable to cold weather, loss of light for navigation, and the failure of many plants to flower and produce nectar. Later, they could be disoriented by increased ultraviolet light, which is visible to many insects.

As normal food supplies for many organisms in natural ecosystems vanished, surviving animals would soon devour whatever remained that could provide nourishment, thereby destroying seed banks, seedlings, eggs, and young animals—the wellsprings of recovery. Insects and other hungry animals would also naturally turn to the most abundant source of food in target regions: crops, which probably could no longer be protected by pesticides. And human survivors, whose stored food and crops were damaged beyond use and undefendable against pests, would turn to natural ecosystems to hunt and gather whatever they could find.

The organisms that would most easily survive such difficult conditions would be the "opportunistic" species that people generally regard as pests or weeds. The combined effects of decimated seedbanks and animal populations, soils degraded by erosion, and continued abnormal weather would retard regeneration of ecosystems. The plant and animal communities that eventually appeared would be impoverished in species diversity and often dominated by undesirable opportunists—very different from the natural communities of coevolved organisms they replaced.

The typical consequences of destroying natural ecosystems and impairing their services to humanity would follow a nuclear war, but on a grander scale. These would include accentuated floods and droughts (in part depending on the disturbance of rainfall patterns induced by nuclear clouds), increased soil erosion, silting of streams and reservoirs, outbreaks of pests and diseases, loss of pollination services, accumulation of wastes, and reduced crop yields. The ability of nature to provide sustenance and absorb abuse would be substantially reduced, just when human survivors were in dire need.

Consequences for Agriculture. During the growing season, crops in fields would be as vulnerable as other annual plants, if not more so, to reduced light and temperatures. Crops would also be more exposed to radioactivity, air pollutants, and UV-B because of proximity to target areas and lack of even minimal sheltering by trees and other plants such as occurs in a complex ecosystem. Relatively exposed cropland soils also lose moisture more rapidly in drought conditions.

The combined assaults in spring or summer of chills, freezes, and wildly fluctuating weather; lack of sunlight; radiation; air pollution; irregular rainfall (too little or too much at once, laden with pollutants and radioactive particles); and UV-B radiation would largely cancel grain, vegetable, and fruit harvests that year.[32]

Even after an autumn or winter war, the next year's harvests could be more or less substantially reduced, depending on the extent to which abnormal weather, radiation, and toxic pollutants lingered into spring and summer. Average summer temperatures in northern mid-latitude continental regions only 2° or 3° C below normal would both shorten the frost-free growing season and prolong maturation of crops enough to shrink harvests sharply. Where growing seasons are already marginal (as in Canada, northern Europe, and most of the Soviet Union's grain belt), crops would fail. Losses would occur even if none of the other effects was a factor and farmers were fully able to plant and manage crops—two big "ifs."

Livestock that escaped the immediate effects of the attack and short-term exposures to radiation, pollutants, and severe weather would be at risk of starvation. Rangeland grasses would be vulnerable to climatic disruption, and feed crops in fields and stored fodder would probably be damaged or destroyed as well. Moreover, many livestock species (especially poultry and pigs) require fairly intensive and constant care, which very likely would not be available.

Modern agriculture's dependence on regular supplies of fuel, fertilizers, pesticides, and other inputs would become a factor in production, both in the short and long term. The disruption of society and its normal transport system (to say nothing of imports) would have serious impacts in industrial nations both on farm production and on the distribution of any successfully harvested food. Also, exposure of crops and seed stores to both ionizing and UV-B radiation could result in less-dependable crop yields and livestock reproduction in later years.

Indeed, the outlook for agriculture in later years is for, at best, a far less stable and predictable regime than prevails today. If unstable weather persisted into the second and third years, crop failures would continue to plague many regions. Without a functional transport system to move any surpluses to areas of shortage, maldistribution of supplies would compound the overall problems of shortage.

Consequences for Civilization

The immediate consequences for societies of combatant nations would depend rather directly on the scale of the war, the level of destruction of cities and transport systems, and the number of people in or near target areas.[33] Even if only key military sites of a major power were targeted in a modest-sized attack, the civilian population would nonetheless suffer millions of prompt casualties from blast, fire, and radioactive fallout. If either superpower were so attacked, 5 to 12 percent of its population—some 12 to 30 million in either case—could be killed by immediate effects, even if cities were not targeted.[34]

Assuming that extensive nuclear fires were not generated and climatic effects were mild at worst, many millions more nonetheless would perish subsequently from exposure, diseases, and cancers induced by sublethal doses of radiation.[35] With so many casualties, a crippled and overwhelmed medical structure, and extensive damage to transport, communications, and industrial infrastructure, if not to habitation, recovery would be extremely difficult, probably requiring decades.[36] If retaliation resulted and other nations were attacked, the capacity for recovery would be even further reduced, because assistance from outside would be unavailable and the probability of nuclear winter consequences would be much greater.

In the event of a full-scale nuclear war involving many nations and attacks on civilian populations, hope for any sort of recovery of targeted regions is dim. External aid would be precluded by disruption of international trade and communications and the spread of severe weather conditions to noncombatant nations, very possibly including the Southern Hemisphere.[37] Hundreds of millions of people would be killed immediately or within a few weeks by radiation sickness, exposure, disease, shock, or a combination; medical care for survivors would be minimal at best.[38] Millions of survivors would be homeless; normal services (electricity, heat, clean water, sanitation) would not exist; weather would be malign and unpredictable; and usable food supplies would quickly run out. Social breakdown in target nations seems likely.[39]

Longer-term food supplies would depend on the amount of destruction inflicted directly on farming regions, the timing of the war and extent of damage to crops in the field or in storage, and survival of a transport system to take materials to farmers and farm products to processors and markets. The food systems of developed nations have so many weak links that providing adequate food for even sharply reduced populations in urban areas after a nuclear attack is extremely dubious. Assuming that remaining food supplies could somehow be distributed and farmers could carry on as usual, nuclear winter conditions nevertheless would cause widespread failure of the next harvest.[40] The likeliest fate of a war survivor after the first month is starvation.

Indeed, starvation seems a probable fate for people far removed from the scene of combat, as spreading nuclear clouds bring disastrous weather patterns to their farms as well. Early deaths caused by a large-scale nuclear war might be well over a billion—one-fifth or more of the world's population. The world's food harvest, of which most surpluses now are produced in the very nations possessing nuclear weapons and likely to fall victim to them, could be cut by more than half in the first year by the climatic effects of such a war. With international food trade arrested, the inability to distribute what was harvested, spoilage and losses to uncontrolled pests, and radioactive contamination would combine to reduce supplies even further. Subsequent harvests would remain low by current standards because of degradation or ruination of farmland by the war itself, destruction of adjacent natural ecosystems and their agriculture-support functions, enhanced UV-B, and persisting adverse weather.

With the major industrial powers in ruins and their populations devastated, it is improbable that industrial civilization in anything like its present form could ever be restored—in large part because the readily accessible and abundant supplies of mineral resources that allowed its original creation would no longer be available. Much of the technology on which that civilization was based would also be lost, along with many of the people who knew how to use it.

The inevitably severe damage to natural and agricultural ecosystems caused by the war and its climatic sequelae would continue to undermine any possibility of rebuilding a world population of several billions; a continued population decline at least for many decades seems more likely. Survivors would be best off in the Southern Hemisphere, but their lives would hardly be easy. Subsequent generations would inherit a world impoverished biotically—with reduced capacity for agricultural production, many formerly productive areas more or less permanently uninhabitable, and a shortage of resources of all kinds. People would carry a burden of lowered resistance to diseases and increased cancers and birth defects. In short, the outlook for humanity's future would be much bleaker than it is today.

NOTES

1. Early works postulating atmospheric effects and other consequences for living systems include T. Tonier, *Nuclear Disaster*, Meridian, Cleveland, 1963; P.R. Ehrlich, "Population Control or Hobson's Choice," in L.R. Taylor, ed., *The Optimum Population for Britain*, Academic Press, London, 1970 (p. 154); P.R. Ehrlich and A.H. Ehrlich, *Population, Resources, Environment*, W.H. Freeman & Co., San Francisco, 1970 (pp. 192–93).
2. P.J. Crutzen and J.W. Birks, "The Atmosphere After a Nuclear War: Darkness at Noon," *Ambio* **11** (1982) 114–25; R.P. Turco, O.B. Toon, T.P. Ackerman, J.B. Pollack, and C. Sagan, "Nuclear Winter: Global Consequences of Multiple Nuclear Explosions," *Science* **222** (1983) 1283–92.
3. See notes 1 and 2 above.
4. U.S. National Research Council, *The Effects on the Atmosphere of a Major Nuclear Exchange*, Washington, D.C.: National Academy Press, 1985.
5. A.B. Pittock, T.P. Ackerman, P.J. Crutzen, M.C. MacCracken, C.S. Shapiro, and R.P. Turco, *Environmental Consequences of Nuclear War: Volume 1. Physical and Atmospheric Effects*, SCOPE 28, Chichester and New York: John Wiley & Sons, 1986.
6. "From a United Nations Study: The Climatic and Other Global Effects of Nuclear War," *Environment* 30, No. 5 (June 1988), 42–45.
7. R.L. Dod, N.J. Brown, F.W. Mowrer, T. Novakov, and R.B. Williamson, "Smoke Emission Factors from Medium Scale Fires: Part 2," *Aerosol Science and Technology* (1989) **10**, 20–27.

8. Ibid.

9. Ibid.

10. R.D. Small and B.W. Bush, "Smoke Produced by Nonurban Target-Area Fires Following a Nuclear Exchange," Pacific Sierra Research Corporation Report No. 1515, Los Angeles, 1985.

11. R.D. Small, B.W. Bush, and M.A. Dore, "Initial Smoke Distributions for Nuclear Winter Calculations," *Aerosol Science and Technology* (1989) **10**, 37–50.

12. "Reference Scenario: How a Nuclear War Might Be Fought," *Ambio* **11** (1982) 94–9.

13. U.S. National Research Council, op. cit.

14. R.D. Small, B.W. Bush, and M.A. Dore, "Initial Smoke Distributions for Nuclear Winter Calculations," *Aerosol Science and Technology*, in press.

15. S.J. Ghan, M.C. MacCracken, and J.J. Walton, "The Climatic Response to Large Atmospheric Smoke Injections: Sensitivity Studies with a Tropospheric General Circulation Model," *Journal of Geophysical Research* (1988) **93**, 8315–37.

16. S.H. Schneider, private communication, 1989.

17. G.A. Glatzmaier and R.C. Malone, "Global Climate Simulations of the ENUWAR Case Studies," Paper MO.16.88, presented at SCOPE ENUWAR Moscow Workshop, March 21–25, 1988.

18. S.J. Ghan, M.C. MacCracken, and J.J. Walton, "The Climatic Response to Large Atmospheric Smoke Injections: Sensitivity Studies with a Tropospheric General Circulation Model," *Journal of Geophysical Research*, in press.

19. H.M. Foley and M.A. Ruderman, "Stratospheric NO Production from Past Nuclear Explosions," *Journal of Geophysical Research* 78 (1973) 4441–51.

20. Ibid. Also H. Johnston, G. Whitten, and J. Birks, "Effect of Nuclear Explosions on Stratospheric Nitric Oxide and Ozone," *Journal of Geophysical Research* 78 (1973) 6107–35.

21. U.S. National Research Council, *Long-Term Worldwide Effects of Multiple Nuclear Weapons Detonations*, Washington, D.C.: National Academy Press, 1975.

22. Glatzmaier and Malone, op. cit.; R.K.R. Vupputuri, "The Effect of Ozone Photochemistry on Atmospheric and Surface Temperature

Changes due to Large Atmospheric Injections of Smoke and NO_x by a Large-Scale Nuclear War," *Atmospheric Environment* 20 (1986) 665–80; S.L. Thompson and P.J. Crutzen, "GCM Simulations of Nuclear War Effects on Stratospheric Ozone," Paper presented at Defense Nuclear Agency Workshop, Santa Barbara, CA, April 19–21, 1988.

23. See note 16.
24. R. Clarke, *London Under Attack; The Report of the Greater London Area War Risk Study*, London and New York: Basil Blackwell, 1986.
25. P.R. Ehrlich et al., "Long-Term Biological Consequences of Nuclear War." *Science* 222:1293–1300, 23 Dec. 1983; P.R. Ehrlich, C. Sagan, D. Kennedy, and W.O. Roberts, *The Cold and the Dark*. New York: W.W. Norton, 1984.
26. M.A. Harwell and T.C. Hutchinson, *Environmental Consequences of Nuclear War: Volume II—Ecological and Agricultural Effects*, SCOPE 28, Chichester and New York: John Wiley & Sons, 1985; M.A. Harwell, *Nuclear Winter: The Human and Environmental Consequences of Nuclear War*. New York: Springer-Verlag, 1984. See also, M.A. Harwell and A.C. Freeman, "Initiating National Case Studies," *Environment* 30, 5 (June 1988), 25–30.
27. See note 26 above.
28. G.M. Woodwell, "The Biotic Effects of Ionizing Radiation," *Ambio*, 11 (1982) 143–48.
29. C.S. Shapiro, T.F. Harvey, and K.R. Peterson, "Radioactive Fallout," in F. Solomon and R.Q. Marston, eds., *The Medical Implications of Nuclear War*, Washington, D.C.: National Academy Press, 1986.
30. See note 26 above.
31. A.H. Ehrlich, "Nuclear Winter: Is Rehabilitation Possible?" in J. Cairns, Jr. ed., *Rehabilitating Damaged Ecosystems*, Boca Raton, CRC Press, 1988.
32. See notes 6, 25, and 26 above.
33. See notes 12 and 24 above.
34. F.N. Von Hippel, B.G. Levi, T.A. Postol, and W.H. Daugherty, "Civilian Casualties from Counterforce Attacks," *Scientific American*, September 1988, 36–42.
35. See note 29 above.
36. F. Solomon and R.Q. Marston, eds. *The Medical Implications of Nuclear War*, Washington D.C.: National Academy Press, 1986; See also note 24 above.

37. W. Green, "New Zealand after Nuclear War," *Environment* 30, 5 (June 1988), p. 28.
38. Solomon and Marston, op. cit.
39. See note 24 above.
40. See notes 6 and 26 above.

8

Cleanup or Buildup: Nuclear Weapons Production in the 21ST Century

DAN W. REICHER

JASON SALZMAN

THE UNITED STATES HAS ARRIVED AT A CROSSROADS IN ITS 40-YEAR HISTORY OF nuclear weapons production. Many of the aged production facilities operated by the Department of Energy (DOE) are in disrepair and a number are shut down because of safety problems. Virtually all the plants are plagued with massive chemical and radioactive waste contamination. The costs of cleanup have been estimated to exceed $100 billion. At the same time DOE is proposing to spend about $50 billion on construction, relocation, and renovation of weapons production facilities.

For years, Congress and the President paid scant attention to the nuclear weapons production facilities. Today, however, the problems in the complex and DOE's plans to address them are facing unprecedented scrutiny. At a time of massive budget deficits as well as improving relations with the Soviet Union, Congress and the President are being forced to weigh the relative importance of nuclear warhead production and protection of human health and the environment. The decisions made by the federal government over the next few years will determine the course of nuclear weapons production for decades to come.

DOE WEAPONS PRODUCTION PLANS

The year 1988 marked a watershed for problems in the U.S. nuclear weapons production complex. Headlines in newspapers and lead stories on the nightly news finally focused public attention on the massive safety and environmental problems that have festered for decades at the 17 sites that make up the complex. A string of reports by DOE and the General Accounting Office (GAO) made it clear that the cost of fixing the problems could exceed $150 billion—roughly equivalent to what it will cost to bail out the savings and loan industry, the year's other crisis.

In December 1988, DOE released its long-awaited plan for rebuilding and cleaning up the weapons complex. In its Modernization Report, DOE proposes to spend almost $50 billion between 1990 and 2010 to convert "the Complex of today into the desired modernized Complex of the future."[1] The Department's scheme includes the construction, relocation, and/or renovation of a number of weapons production plants. The plan, however, devotes far less money—less than half of total estimated needs—to the cleanup of serious contamination across the complex. As a result, significant threats to human health and the environment will remain unaddressed in 2010, when DOE expects to have rebuilt U.S. warhead production capability. In heading down this path, DOE seems poised to repeat many of the mistakes that created the current crisis.

New Production Facilities

New Production Reactors. The most expensive new initiative is the proposed construction of two new production reactors for the manufacture of tritium and, if needed, plutonium for nuclear weapons. Development of these new production reactors (NPR) would likely take over ten years and cost more than $10 billion.

DOE justifies construction of both NPRs primarily on the need for tritium, which, because of its 12.3 year half-life, is regularly replenished in warheads. In recent years tritium has been produced at the Department's problem-plagued Savannah River Site (SRS) reactors (as of April 1, 1989, with Westinghouse's takeover, the facility's name was changed from Savannah River Plant to SRS). The SRS reactors were shut down in mid-1988 because of safety problems.

The Energy Department has selected the SRS as the preferred location for "urgent" construction of a new heavy water reactor, and the Idaho National Engineering Laboratory for construction of four modular high-temperature gas-cooled reactors. Two alternatives—construction of a new liquid metal reactor or conversion of a partially completed commercial light water reactor in Washington State—were rejected by DOE, after a review by the Department's Energy Research Advisory Board.

Under DOE's proposal, which has not yet been approved by Congress and the President, the new South Carolina reactor would provide 100 percent of U.S. tritium requirements, and the four Idaho reactors would provide an additional 50 percent capacity for tritium and plutonium production. Members of Congress and independent experts increasingly question the need for the Idaho reactors. However, since the reactors would also produce electricity, they are being promoted in some quarters as a way to demonstrate the viability of gas-cooled reactors for the next generation of commercial nuclear power plants.

Special Isotope Separation (SIS) Plutonium Plant. DOE hopes to begin construction in the near future of the first new U.S. nuclear weapons plant in 25 years at the Idaho National Engineering Laboratory, near Idaho Falls. The Special Isotope Separation (SIS) facility—costing $3 billion or more to develop, build, and operate—would use lasers to convert a limited amount of DOE-owned fuel-grade plutonium to weapon grade, a job that would take about eight to ten years to complete.[2]

The Department, however, has not explained how the SIS plant would be used during the remainder of its 30-year operational life. In the early 1980s, DOE's major argument for the development of the SIS technology was to enrich plutonium extracted from commercial spent nuclear fuel for use in nuclear weapons. Such an application would effectively transform U.S. nuclear power plants into weapons materials factories.[3] It would also run counter to our nation's efforts to stem the spread of nuclear weapons to other nations by, in effect, demonstrating a technology to produce warhead materials from commercial spent fuel. However, DOE now states that the SIS plant will not be used for this purpose.

In 1988, the House and Senate unanimously passed amendments to the 1989 defense authorization bill delaying the construction of SIS. Serious concerns were raised by members of the Armed Services Committees and environmental and arms control groups about the need for the facility and whether the underlying technology was ready for deployment. The safety of SIS was also a significant issue. In an accident, plutonium vaporized in the SIS plant would burn spontaneously if exposed to air.

In 1989, the GAO suggested that Congress reevaluate the SIS project. Subsequently, Congress cut DOE's requested funding for the Idaho plant by over 60% and refused to allow construction to begin. Prospects are good that development of the plant will be halted entirely in 1990. (Editor's note: President Bush officially halted development of the SIS plant in January, 1990 with the release of his proposed Fiscal Year 1991 budget.)

Relocation of Facilities

In its Modernization Report, DOE proposes to halt or curtail activities at a number of DOE weapons plants. According to the report, the Fernald facility in Ohio would be closed entirely. Built in 1953, the plant has produced uranium target rods for plutonium production. According to DOE, Fernald is "severely deteriorated and the projected workload is decreasing."[4] In light of these conditions, DOE states that it would be "cost effective" to transfer its remaining functions to an as yet unnamed DOE site in the early 1990s.

By the year 2000, DOE proposes to halt all nuclear weapons production activities at the Hanford facility, in Washington State, which has been in operation since the Manhattan Project. However, because of the extent of radioactive and chemical contamination, extensive cleanup activities will have to continue at Hanford well into the next century.

The Department also proposes to move operations from the Rocky Flats plant in Colorado to other facilities over the next several decades at a cost of over $5 billion. DOE has proposed relocation of the facility—which has produced plutonium components for weapons since 1951—in part because the "public perception of risk is such that even insignificant incidents involving nuclear materials may result in major impacts on production."[5] According to DOE:

> Active environmental, antinuclear, and disarmament groups are using available information on operations of the Complex to convince the public that such operations are too risky and, in most cases, unnecessary. . . . Consequently, there exists a concern by many in our society that operation of some of DOE's nuclear weapons facilities involves unacceptable levels of risk.

> The Rocky Flats plant near Denver will likely continue to face a problem of public opposition, due to the increasing population in areas surrounding the site, in spite of any DOE attempts to modernize this facility in the future.[6]

Possible relocation sites include the Savannah River plant in South Carolina, the Idaho National Engineering Laboratory, and the Pantex plant in Texas.

DOE also proposes to transfer to as yet unspecified locations all nuclear operations from the Mound facility in Ohio, where detonators and various parts of the firing circuits for nuclear weapons are produced. Mound's non-nuclear activities would continue, including the production of non-nuclear explosives.

Renovation of Facilities

The Department is proposing major renovations to uranium enrichment facilities at the Y-12 plant in Tennessee, and processing plants at the Savannah River Site and Rocky Flats (despite the proposed relocation). DOE has already begun major upgrades to the Savannah River reactors that will cost several hundred million dollars. Additionally, Congress has begun to fund a number of programs for general repair of equipment and facilities across the complex.

The Bill for New Weapons Production Activities

DOE has estimated that the total cost for construction, relocation, and renovation initiatives will be about $50 billion. It is likely, however, that the actual funding requirements will be much higher, given DOE's track record in predicting costs of large projects. In 1988, the House Appropriations Subcommittee on Energy and Water Development determined that 21 major DOE construction projects had, on average, increased in cost by nearly 500 percent from the time they were initiated until they were completed or abandoned. The subcommittee concluded that DOE's estimates for new construction projects "must be challenged in light of the Department's past competence in estimating the cost of and managing the construction of major projects."[7]

DOE CLEANUP PROGRAMS

Dwarfing DOE's ambitious production initiatives are the plans for cleanup of massive quantities of radioactive and hazardous wastes that have contaminated groundwater and soil at most of the weapons facilities. DOE's proposals to address environmental problems can be broken down into four broad programs:

- A "remedial action" program to clean up more than 3,000 sites at weapons plants where solid, hazardous, and/or radioactive wastes have been dumped or leaked;
- A "decontamination and decommissioning" program to clean up and, in some cases, dismantle and dispose of buildings that are no longer in use;
- A "radioactive waste management" program for construction of permanent waste repositories and waste processing plants;
- An annual "base program" to cover day-to-day activities to comply with environmental, safety, and health standards, including sampling and analysis, operation and maintenance, personnel training, technology development, audits, and inspections.

DOE's Cleanup Costs

The Energy Department estimates that the total cost of cleaning up waste sites, decontaminating buildings, and disposing of waste will be $42 billion to $72 billion, plus an additional $1.4 billion annually to cover DOE's base program.[8] GAO estimates that these costs could be higher: $80 to $110 billion.[9]

The actual cost of cleaning up the weapons complex could be still more. First, the nature and extent of many cleanup problems, as well as their solutions, are unclear, making estimates tentative at best. For example, the Environmental Protection Agency's (EPA) estimate of the cost to clean up a commercial Superfund site has more than doubled in five years, from an average of $8.4 million in 1984 to $20 million today. Second, the cleanup problems at many waste sites worsen each day as contaminants continue to migrate into soil and water. Third, according to GAO, some sites may require "long-term institutional care" because of the difficulty of remediation. Costs could accrue essentially forever in these so-called "national sacrifice zones." Finally, Congress, EPA, and the states may strengthen environmental laws and regulations over the next decade, increasing cleanup costs further.

Cleanup Programs		**Estimated Cost (billions of dollars)**	
		DOE	**GAO**
Remedial Action		35–63	35–65
Radioactive Waste Management		4–5	30
Decontamination/Decommissioning		3–4	15
	TOTAL*	42–72	80–110
Base Program (annual cost)		1.4	____**

* The DOE total does not include $6 billion to $14 billion that the Department estimates will be needed to bring existing facilities into compliance with environmental standards. The GAO total does not include the $20 billion the agency estimates the same task will require.

** GAO does not include base program funds in its estimate because it considers them a "normal cost of doing business that would be necessary whether DOE had any problems or not."

SOURCE: General Accounting Office (GAO/RCED-88-197BR), July 1988, and Department of Energy (DOE/EH-0079), December 1988.

Table 1: Department of Energy Cleanup Costs

DOE's Cleanup Costs in Perspective

Whatever the precise figure, DOE's bill for cleanup is staggering. Even if the cost is half of the lowest current estimate, it would still make small change of "expensive" federal projects like the $6 billion Superconducting Supercollider. And if the cleanup bill is $100 billion—a reasonable estimate based on DOE and GAO figures—it would be more than the total amount spent to put a man on the moon in the Apollo space program. Adding the $50 billion that DOE estimates will be required for modernization of the complex would result in a bill requiring more than $1,600 from each U.S. household. If costs escalated to $250 billion, they would be about equivalent to the nation's total spending on nuclear warhead production from World War II to the present. And if the total bill reached $500 billion, it would exceed what it cost the United States to fight the Vietnam War.

ARE ADEQUATE FUNDS AVAILABLE?

The GAO recently reported that DOE's bill for modernization and cleanup of the nuclear weapons production complex is "one of the major areas of explosive unfunded costs that will have to be dealt with at the same time as the budget deficit." And DOE itself recognizes that future funding for weapons production and cleanup initiatives will be limited. In the Modernization Report, the Department admits: "[A]ccomplishing all the recommendations of the Plan in the timeframe specified will require resources greater than those now expected to be available in future years."[10]

In December 1988, then DOE Secretary John Herrington wrote to White House Chief of Staff Kenneth Duberstein seeking more money for DOE environmental programs. Herrington's letter illustrates not only how

Superconducting Supercollider[b]	$ 5 billion
Washington, D.C. Subway[c]	$ 11 billion
Apollo Space Project[d]	$ 84 billion
Cleanup and Modernization of the Nuclear Weapons Production Complex	**$150 billion??**
Nuclear Weapons Production WWII to Present[e]	$260 billion
Vietnam War[f]	$480 billion
All Federal Highways[g]	$920 billion

[a] All figures in 1989 dollars.

[b] *The Facts on File World Political Almanac,* C. Cook, ed., New York: Facts on File, 1988.

[c] Metro Office of Public Affairs, Washington, D.C.

[d] NASA Office of Public Affairs, Washington, D.C.

[e] Updated from Cochran et al., *Nuclear Weapons Data Book,* Vol. II, Cambridge, Mass.: Ballinger Publishing Co., 1987, p. 2.

[f] *The People's Almanac No. 2,* D. Wallechinsky and I. Wallace, eds., New York: W. Morrow and Co., 1978.

[g] M.H. Best and W.E. Connolly, *The Political Economy,* 2nd ed., Lexington, Mass: Heath, 1982; *Statistical Abstract of the United States,* Washington, D.C.: Government Printing Office, 1988.

Table 2: How Much Things Cost[a]

tight DOE's budget is, but also how decisions about funding for weapons production and environmental cleanup are increasingly linked:

> I must in good conscience advise you that . . . the [weapons production] system will, in all likelihood, be operating beyond the bounds of acceptable risk. . . . If available funding is used for production, then unresolved environmental and safety issues are likely to lead to further shutdowns in the weapons complex. If available funds are devoted to environment and safety, then production goals will not be satisfied.[11]

Duberstein refused to grant the increase, directing Herrington instead to make up the shortfall from DOE's own budget.

Over the long run, even if Congress fully funds DOE's plans as outlined in the Modernization Report and costs do not increase, after two decades the nation would still face a severely contaminated weapons complex. Under the spending proposal made in the Modernization Report, DOE would spend about $50 billion over 20 years to build, relocate, and renovate weapons plants. In contrast, just $32 billion would be spent to clean up waste sites and abandoned buildings over the same period. Assuming cleanup of DOE waste sites and buildings will cost $65 billion and the Department spends $32 billion on these activities over the next two decades, more than half the cleanup job would still remain in the year 2010. And if the total bill is $100 billion, cleanup would be only about one-third complete by this date. DOE's emphasis on production over cleanup will mean that massive amounts of contamination will remain in soil and groundwater well into the next century, posing a significant threat to human health and the environment.

NUCLEAR WEAPONS PRODUCTION AND ARMS CONTROL

DOE's proposal to modernize the weapons complex comes at a time when several hundred nuclear warheads are being withdrawn from Europe under the Intermediate Nuclear Forces (INF) Treaty, and some 4,000 U.S. warheads may be retired under the Strategic Arms Reduction Treaty (START) presently being negotiated.

In light of these arms control initiatives, as well as serious cleanup needs and daunting fiscal constraints, members of Congress have begun to question whether the United States should commit tens of billions of dollars to building a new generation of nuclear weapons production facilities and renovating and relocating existing ones. The debate has focused on whether any new plutonium facilities are needed, and the urgency of additional tritium production capacity.

"Awash in Plutonium"

With a radioactive half-life of 24,000 years, plutonium-239 essentially lasts forever and is routinely recycled. In fact, most plutonium for new warheads is now obtained from the retirement of old weapons and not from new production. Greatly accelerated production rates during the Reagan administration, in addition to recent large-scale retirements, have created a plutonium surplus. In fact, DOE Secretary Herrington testified before a congressional subcommittee in 1988 that "[w]e're awash in plutonium. We have more plutonium than we need."[12]

The accelerated plutonium production rates during the Reagan administration were justified by the Nuclear Weapons Stockpile Memorandum, a classified document produced annually by the departments of Defense and Energy outlining U.S. requirements for nuclear materials. In 1987, Senator Mark Hatfield (R-Ore.) analyzed the adequacy of the Stockpile Memorandum as a guide for producing plutonium. He concluded that because the document is drafted annually without congressional input, it is "a wish list of nuclear weapons for the Pentagon."[13] According to Hatfield, from 1982 to 1987 the Stockpile Memorandum overprojected—sometimes by a factor of two—the number of warheads that were actually built. DOE nevertheless based its production of warhead materials on the inflated warhead projections, resulting in a surplus of plutonium.

Arms Control Agreements and Plutonium Requirements

Arms control agreements will swell this plutonium surplus. Warheads from the approximately 500 weapons retired under the INF Treaty are being stored, while the delivery systems are being dismantled. The warheads or the nuclear materials they contain will be available, if needed, for the construction of new weapons. The same would likely be true of the estimated 4,000 warheads that would be retired under a START agreement.

However, DOE contends that the negotiation of arms control treaties like START should not affect plans to produce nuclear materials, including plutonium. DOE's Modernization Report states that "under most nuclear arms control scenarios, the capabilities of the Complex, both technical and productive, must be maintained," and that the complex must "provide a capability for responding to a possible breakout from any arms control agreement."[14] Nonetheless, DOE's Acting Assistant Secretary for Defense Programs, Troy Wade, admitted in 1988 that arms control could have an impact on the need for plutonium from the proposed SIS plutonium plant. Wade stated, "Should there be a START agreement and perhaps follow-on agreements where substantial weapons are returned, clearly it [the SIS plutonium plant] would be worth revisiting."[15]

In addition to plutonium that will become available from the INF Treaty and other potential arms control agreements, large amounts of plutonium can also be obtained from the estimated ten metric tons of plutonium scrap left over from previous nuclear weapons production.

Bilateral Plutonium Cutoff

With such a large amount of plutonium on hand, many experts, including former CIA Director William Colby, former chief of the SALT I Delegation Gerard Smith, and former Director of the Arms Control and Disarmament Agency Paul Warnke, have called on Congress and the President to declare a two-year moratorium on U.S. plutonium production and challenge the Soviet Union to negotiate a permanent, bilateral halt to the production of plutonium—and highly enriched uranium—for nuclear weapons.[16] DOE Undersecretary Joseph F. Salgado conceded in 1988 that such a two-year moratorium "would not have a negative impact"[17] on national defense.

On a number of occasions in the 1960s, the United States proposed a bilateral cutoff of the production of plutonium and highly enriched uranium for nuclear weapons to the Soviet Union, but the Soviets rejected the offer. Nevertheless, in 1964 President Johnson unilaterally halted production of highly enriched uranium for warheads, and the United States has not produced any since.[18]

In 1982, the Soviets announced their support of a bilateral cutoff. Foreign Minister Andrei Gromyko stated that "one of the initial stages" in a disarmament program could be the "cessation of production of fissionable materials for manufacturing nuclear weapons."[19] However, the United States did not support the proposal.

In April 1989, Soviet Secretary Gorbachev announced that the Soviet

Union would halt the production of highly enriched uranium for nuclear weapons in 1989, and by 1991 would shut down two reactors that currently produce weapon-grade plutonium, in addition to one closed in 1987. According to Gorbachev, no new reactors would be commissioned as replacements. Gorbachev called the initiative "a major step toward the complete cessation of production of fissionable materials for use in weapons."[20]

Officials from the Bush administration downplayed the significance of the Soviet initiative, arguing that the Soviets have several other reactors that produce plutonium. They also expressed overall opposition to constraints on nuclear materials production.

In May 1989, however, Senators Edward Kennedy (D-Mass.) and Timothy Wirth (D-Colo.) and Representatives Ron Wyden (D-Ore.) and Dante Fascell (D-Fla.), along with 100 cosponsors, introduced the International Plutonium Control Act, which would encourage U.S. negotiations with the Soviets for a bilateral, verifiable agreement to halt permanently the production of additional plutonium and highly enriched uranium for warheads.

Backers of the legislation asserted that such a cutoff would be verifiable. In fact, the United States in 1969 indicated that a cutoff could be verified by independent nuclear material control and accounting at declared nuclear facilities supplemented by surveillance satellites and other intelligence means. On-site inspections of Soviet and U.S. nuclear facilities would further ensure that such an agreement would not be violated.

Legislation similar to that introduced was adopted by a two-to-one margin in the House but unfortunately this helpful measure was not enacted.

Tritium Production Capacity

Unlike plutonium, tritium has a relatively short half-life (12.3 years), and therefore is regularly replenished in warheads.[21] Some independent experts have suggested that the United States and the Soviets agree to halt bilaterally the production of tritium and to use the decay of the remaining tritium supply as a pacing mechanism for further arms reductions. Proponents of this approach—called the "Tritium Factor"—also argue that a halt in tritium production would permit the United States to defer or even cancel plans for the construction of one or more new production reactors. The Tritium Factor has been criticized by experts both inside and outside the government because, among other things, it would be difficult if not impossible to verify, it would likely affect U.S. and Soviet weapon stockpiles differently, and it could produce an intensive effort in both nations to design and test new warheads that rely less on tritium.

CONCLUSION

DOE's approach to nuclear weapons production, as reflected in the Modernization Report, does not recognize two fundamental facts: first, the serious long-term threat posed by contamination in the weapons complex; second, improving relations with the Soviets that may lessen the need for nuclear warhead production capacity. Left to its own devices, DOE may continue to ignore these facts, and as a result, cleanup will continue to take a backseat to production. (Editor's Note: In January, 1990 Energy Secretary James Watkins agreed to prepare Programmatic Environmental Impact Statements on the cleanup and modernization of the weapons complex in the wake of litigation brought by the Natural Resources Defense Council on behalf of twenty-one citizen organizations.)

Sustained public and congressional pressure will be needed to change this situation. The public must demand a comprehensive and public analysis of the future of the weapons complex that takes into account current and potential arms control agreements and the serious health and environmental threats posed by radioactive and chemical contamination.[22] Congress must realign DOE budget priorities to reflect improving relations with the Soviets and the critical need for cleanup. Only then will the Department be on the right track toward accomplishing its mission of protecting this nation's security.

NOTES

1. U.S. Department of Energy, "United States Department of Energy Nuclear Weapons Complex Modernization Report: Report to Congress by the President," Dec. 1988, p. 2.
2. Final Environmental Impact Statement, Special Isotope Separation Project, U.S. Department of Energy, Volume 1, at S-5, 1-1.
3. Dan W. Reicher and Jason Salzman, "High-Tech Protest Against Plutonium Plant," *Bulletin of the Atomic Scientists,* November 1988, p. 23.

4. U.S. Department of Energy, "United States Department of Energy Nuclear Weapons Complex Modernization Report: Report to Congress by the President," Dec. 1988, p. 16.
5. *Id.* at 18.
6. *Id.* at 9.
7. U.S House of Representatives, Appropriations Committee Subcommittee on Energy and Water Development, Report (100-618), May 1988, p. 55.
8. U.S. Department of Energy, "Environment, Safety, and Health Needs of the U.S. Department of Energy, Volume 1: Assessment of Needs" (DOE/EH-0079), Dec. 1988.
9. General Accounting Office, "Dealing With Problems in the Nuclear Weapons Production Complex Expected to Cost Over $100 Billion," GAO/RCED-88-197BR, July 1988. Also see GAO reports GAO/RCED-88-62 and GAO/RCED-86-192.
10. U.S. Department of Energy, "United States Department of Energy Nuclear Weapons Complex Modernization Report: Report to Congress by the President," Dec. 1988, p. iv.
11. *Washington Post*, Dec. 18, 1988, p. A1.
12. John Herrington, Secretary of Energy, Testimony before the House Appropriations Subcommittee on Interior and Related Agencies, Feb. 23, 1988.
13. Senator Mark Hatfield, "The Plutonium Cushion: Report on U.S. Defense Plutonium Needs," October 1987, at 4.
14. U.S. Department of Energy, "United States Department of Energy Nuclear Weapons Complex Modernization Report: Report to Congress by the President," Dec. 1988, p. 6.
15. Troy Wade, Assistant Secretary for Defense Programs, Testimony before the Senate Armed Services Committee, Subcommittee on Procurement and Nuclear Strategy, Feb. 1988.
16. "The Plutonium Challenge: An Open Letter to the President and the Congress of the United States," The Plutonium Challenge Coalition, November 5, 1987.
17. *Washington Post*, Feb. 28, 1988, p. A-4.
18. David Albright and Christopher Paine, "A Case Against Producing Nuclear Material," *Bulletin of the Atomic Scientists*, January/February 1988, p. 47.

19. United Nations General Assembly, June 15, 1982.
20. Craig R. Whitney, "Gorbachev Plans to Stop Producing Uranium for Arms," *New York Times*, April 8, 1989, p. 1.
21. David Albright and Theodore Taylor, "A Little Tritium Goes a Long Way," *Bulletin of the Atomic Scientists*, January/February 1988, p. 39.

PART TWO

Beyond the Atom

9

Weapons Forsworn: Chemical and Biological Weapons

JOHN W. BIRKS

THROUGHOUT HISTORY, CHEMICAL AND BIOLOGICAL WEAPONS HAVE BEEN considered the most repugnant of all means of killing. This, added to their unpredictability, has made these weapons less popular than explosives and incendiaries. Nevertheless, large-scale chemical warfare was waged in World War I, causing more than one million casualties and more than 100,000 deaths; the United States made widespread use of tear gas and defoliants in the Vietnam War; and the use of chemical weapons by Iraq played a major role in bringing Iran to the peace table. The latest development could easily challenge the widely held view that chemical weapons are of little tactical use.

Fortunately, public attention recently has been brought to bear on the dangers of nuclear war. I fear, however, that while our energies are so consumed in working to bring these weapons of mass destruction under control, even more dangerous weapons could be developed. Biology is undergoing a revolution today that can be compared to that which occurred in physics in the 1930s and 1940s. The chemical basis of life has been discovered, and with that discovery has come the ability to modify life forms to suit our purposes—for good or for evil. Our understanding of the binding forces within the atomic nucleus rapidly led to the development of fission and fusion bombs of previously unimagined destructive power. But our understanding of the nucleus of living cells, if misused, has even greater destructive potential. Microorganisms such as bacteria and viruses may be imagined as tiny bombs so small as to be invisible to the human eye. These bombs may be used to kill men, women, and children, or to kill only the livestock and crops their lives depend on. But unlike atomic bombs, these tiny bombs produce no collateral damage: houses, factories, and power plants remain intact as the spoils of war.

The first verified use of biological weapons occurred in 1346, when the Golden Horde of Mongol Tartars catapulted plague-infected cadavers over the walls of Feodosia (then Kaffa), an important port city on the north coast of the Black Sea.[1] The heavily fortified city, which had withstood three years of siege, fell shortly thereafter. The escaping inhabitants apparently introduced the plague to Italy, whence it spread into the rest of Europe and caused the pandemic that ravaged the entire continent from 1348 to 1350. The plague bacterium was also used by Japan against the Chinese in the Second Sino-Japanese War of 1937–45, which merged into World War II; Japan attacked at least 11 cities of China with this and the bacteria causing cholera, anthrax, and typhoid fever.[2] At least 700 Chinese died from plague alone, but the attacks appear to have been of very little military utility.

Chemical weapon stockpiles are known to be maintained by the United States, the Soviet Union, France, and Iraq. In addition, Iran, Egypt, Syria, Libya, Israel, Ethiopia, Myunmar (formerly Burma), China, Taiwan, North Korea, South Korea, Thailand, and Vietnam have been reported to have chemical weapons.[3] No country admits to possessing biological weapons today, although at least three nations have possessed them in the past: Japan in the 1930s and 1940s, the United Kingdom during the 1940s, and the United States from the 1940s through the 1960s.[2] Possession of biological warfare agents by the Soviets has often been alleged.

The use of chemical and bacteriological warfare agents is outlawed by the Geneva Protocol of 1925, an international arms control agreement not ratified by the United States until 1975.[4] Biological warfare agents are further prohibited by the "Convention on the Prohibition of the Development, Production and Stockpiling of Bacteriological (Biological) and Toxin Weapons and on Their Destruction," a treaty ratified by more than 100 nations and heralded as the world's only disarmament treaty. Despite this treaty, biological warfare research in the United States has received large increases in funding over the past few years, presumably for defensive measures only, such as immunizations and the development of other protective measures. A similar global disarmament treaty for chemical weapons was called for in the 1975 convention on biological weapons and is still being negotiated in Geneva. In the meantime, enormous stockpiles of these weapons have been accumulated by the United States, the Soviet Union, France, Iraq, and possibly other nations. The production, storage, transportation, and destruction of these weapons continue to pose enormous risks for the nations that harbor them.

CHEMICAL WEAPONS

A chemical weapon system is composed of four parts: 1) a method to deliver the weapon, 2) munitions to disseminate the chemical agent, 3) the chemical agent and/or its activation system, and 4) the part played by the environment in transporting the disseminated chemical to each target individual.[5] Obviously, the toxicity of the agent is not the only consideration. The chemical agent should be sufficiently stable to be stored for long periods of time. Its persistence in the environment, once dispersed, must be considered; the tactical situation dictates whether a chemical agent of either short-term or long-term stability may be desirable. For this reason, compounds having a range of vapor pressures and varying rates of hydrolysis (reactions with water) have been synthesized. The arsenals contain chemical compounds that are only temporarily debilitating, such as the tear gases, in addition to those that are lethal. The classes of chemical warfare agents and some of the most important examples are given in Table 1.

World War I Vintage Gases. The principal gases used in World War I were chlorine, phosgene, trichloromethylchloroformate (commonly called diphosgene), chloropicrin, hydrogen cyanide, and mustard gas. Chlorine, phosgene, diphosgene, and chloropicrin cause damage to the respiratory system and death by pulmonary edema at high exposures, or, if the exposure is less, bronchitis or pneumonia may result due to bacterial invasion of the damaged tissues.[7] Hydrogen cyanide (HCN) causes death by inhibiting oxygen uses and transport. It blocks the oxidation of acetate and other metabolites within the mitochondria of cells by forming a stable complex with the oxidation enzymes.[8] HCN has long been used to exterminate insects and rodents in the holds of ships, in warehouses, in railway cars, and on fruit trees. It has also been used as a means of human execution. Because of its high volatility, and a density less than that of air, HCN is rapidly dispersed in the atmosphere, and as a result was relatively

Class of Agent	Army Code	Common Name	Chemical Structure	Symptoms
Choking Gas	CG	phosgene	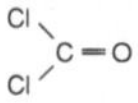	coughing, retching, frothing at mouth, cyanosis; asphyxia; pneumonia
Blood Gas	AC	hydrogen cyanide	HCN	giddiness, convulsions, unconsciousness, asphyxia
Blister Gas	H	mustard gas	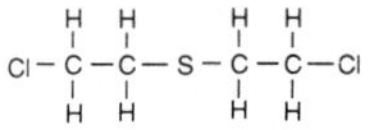	eyes: inflammation, photophobia, ulceration, blindness
"	HN3	nitrogen mustard	$N(CH_2CH_2Cl)_3$	skin: redness, irritation, blisters
Nerve Gas	GA	tabun	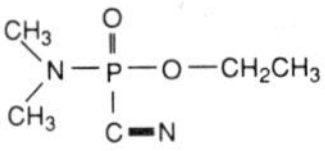	drooling, sweating, nausea, vomiting, cramps, involuntary defecation or urination, twitching, jerking, staggering, headache, confusion, drowsiness, coma, convulsions, asphyxia
"	GB	sarin	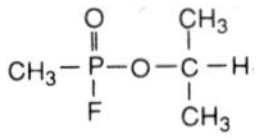	
"	GD	soman	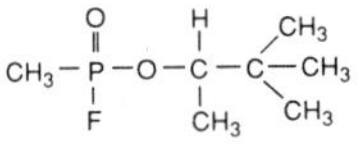	
"	VX		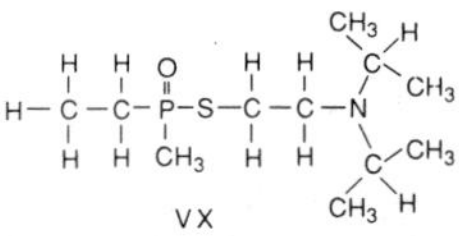 VX	

Table 1: Principal Chemical Warfare Agents

Class of Agent	Army Code	Common Name	Chemical Structure	Symptoms
Tear Gas	CN		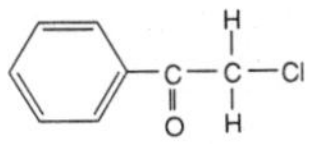	burning feeling on moist skin, copious lacrymation; nonfatal
"	CS		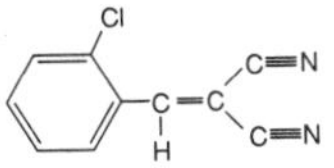	stinging and burning feeling on skin, coughing, tears, chest tightness, nausea; nonfatal
Hallucinogen	BZ		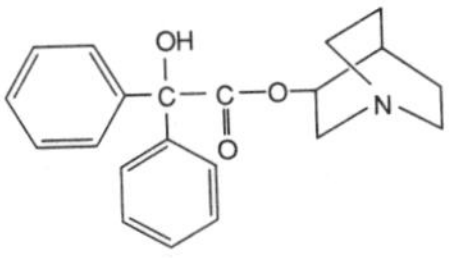	slowing of physical and mental activity, giddiness, disorientation, hallucinations, occasional maniacal behavior

Adapted from ref. 5.

Table 1: Principal Chemical Warfare Agents continued

ineffective when used by the French army in World War I.[6] Nevertheless, the United States stockpiled hydrogen cyanide weapons during World War II.

Mustard gas was developed by the Germans near the end of World War I and proved to be the most effective chemical agent in producing casualties.[6] Mustard gas belongs to the class of chemical warfare agents known as vesicants, or "blister gases."[7] Such compounds react with components of any living cell (by alkylation). Their principal targets are the skin and eyes, which they burn and blister. Large stockpiles of mustard gases were accumulated during World War II but never used. They remain in the U.S. arsenal, but are scheduled for destruction as an obsolete chemical warfare agent.

Nerve Gases. The nerve gases tabun (GA), sarin (GB), and soman (GD) were developed and stockpiled by Germany shortly before and during World War II but never used.[9] These agents inhibit acetylcholine esterase, an enzyme in nerve cells that has the important function of destroying one of the neural transmitters, acetylcholine.[8] In the absence of this enzyme, receptor nerves fire repeatedly, overstimulating muscles, glands, and organs. The heart beats wildly and irregularly, and death results from convulsions. The mode of action is identical in chemically related organophosphorous compounds commonly used as insecticides (e.g., malathion and parathion). Nerve gases have the military advantage of acting very rapidly and requiring much lower concentrations than other chemical agents to cause incapacitation or death. The lethal inhalation exposure to GB, for example, is about 15 times less than mustard gas and about 32 times less than phosgene.[10] Another class of nerve gases, the V-agents, which are even more toxic, were developed by the British in the 1950s. About 80 percent of the U.S. nerve gas in the present arsenal is GB, and about 20 percent is the more deadly agent VX.[11] On a per weight basis, VX is more than 100 times more lethal than GB when in contact with skin and about 3 times more lethal by inhalation.[10] VX has a lower vapor pressure than GB and is much more persistent. In cold weather it may render a region dangerous for up to several weeks. This is an advantage for applications deep in enemy territory but a disadvantage for battlefield uses. For this reason many of the stockpiled weapons containing VX—such as land mines, spray tanks, and artillery shells—are no longer considered viable.

Harassing Agents. Commonly known as tear gas, these compounds have been developed as nonlethal agents for riot control and police actions. Some are particularly irritating to the eyes (lachrymators), some to the inner surfaces of the nose and upper breathing tubes (sternutators), some to the skin (orticants), and some, if swallowed, to the upper reaches of the alimentary canal (vomiting agents).[5] The tear gas now used almost exclusively is CS, which was often employed in Vietnam to force combatants from buildings and underground tunnels. CS is more effective than the older agent CN in that it replaces the latter's lachrymatory action with a more severe skin irritation and is more stable in storage. These agents will not be subject to the chemical weapons treaty currently being negotiated, and as they are lethal only in very high concentrations, they do not pose a particularly great storage and transportation hazard.

Hallucinogens. Another approach to chemical warfare is to disperse mind-altering drugs that will cause enemy personnel to hallucinate or

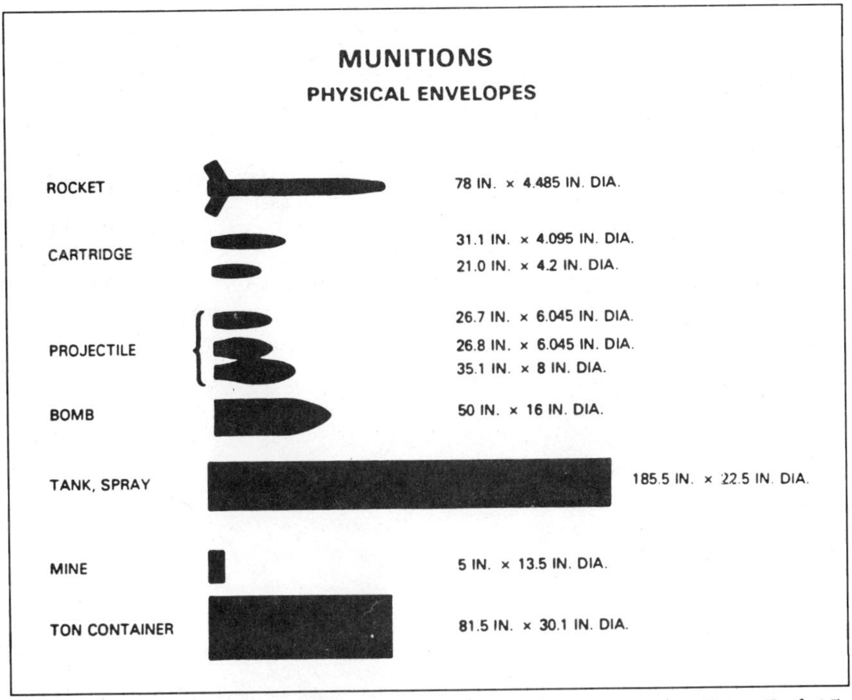

Figure 9-1: Physical dimensions of chemical warfare munitions. From Ref. 15.

become confused. One such agent developed by the United States, BZ, is apparently more potent than the 1960s psychedelic drug of abuse, LSD.[12-14] As discussed below, a small amount of BZ (about 45 tons) was stockpiled at the Pine Bluff Arsenal.[15]

U.S. Arsenal of Chemical Weapons

The U.S. stockpile of chemical agents and munitions consists primarily of M55 rockets (containing GB or VX nerve gases), M23 mines (containing VX), 105-mm, 155-mm, 4.2-inch, and 8-inch projectiles (containing GB, VX, or mustard agent), MC-1 MK94 bombs (containing GB), and 1-ton containers of GB, VX, and mustard.[15] Figure 1 provides the physical dimensions of these weapons, and Figure 2 contains schematic diagrams of the M55 rocket and the M23 land mine.

With the exception of the new binary weapons manufactured in the past two years, all chemical agents and munitions are at least 20 years old, and some are 45 years old or older.[15] Mustard gas was manufactured during World War II and again in 1954–55. It has been estimated that there is a total of 40,000 tons of nerve gases, consisting of approximately 80 percent GB and 20 percent VX in the current arsenal.[11]

The hallucinogen BZ was manufactured at Pine Bluff Arsenal in 1963–64.[15] Some 40 tons of the white crystalline powder were mixed with a pyrotechnic and loaded into two types of cluster bombs. Another 5 tons were placed in bulk storage along with 600 tons of BZ-contaminated residue from the manufacturing process and an undisclosed amount of residue from a 1971 fire in a storage igloo. The relatively small amounts of this agent were scheduled for disposal in 1987–88.

The chemical munitions are stored on the Johnston Atoll near Hawaii in addition to eight army depots in the continental United States:

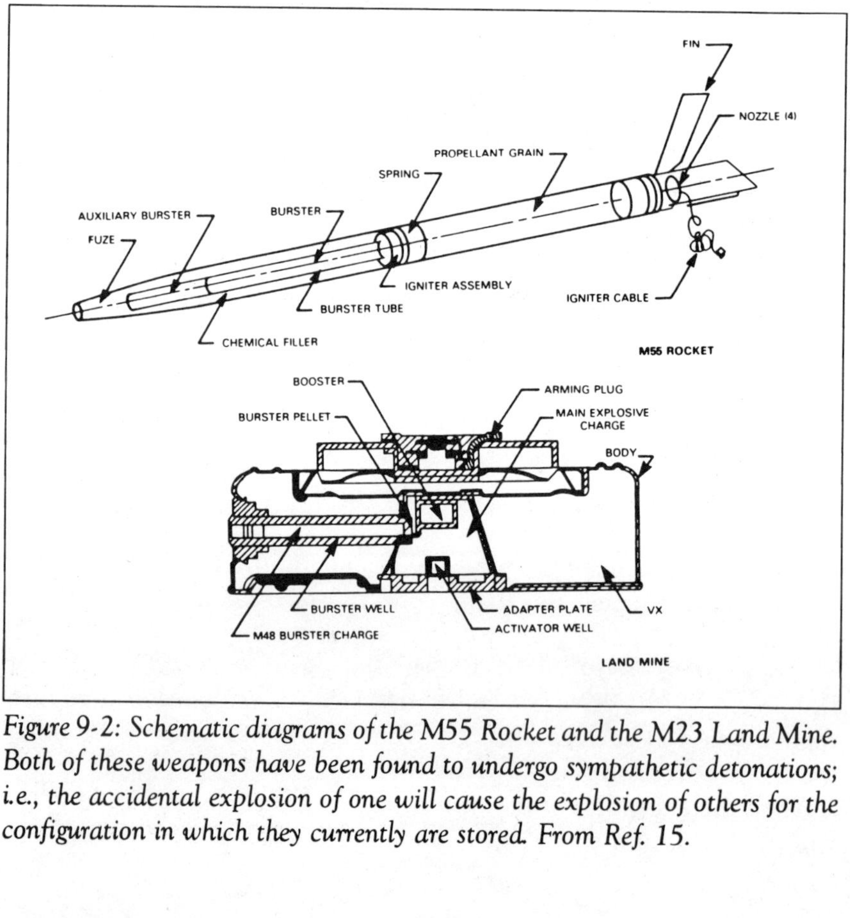

Figure 9-2: Schematic diagrams of the M55 Rocket and the M23 Land Mine. Both of these weapons have been found to undergo sympathetic detonations; i.e., the accidental explosion of one will cause the explosion of others for the configuration in which they currently are stored. From Ref. 15.

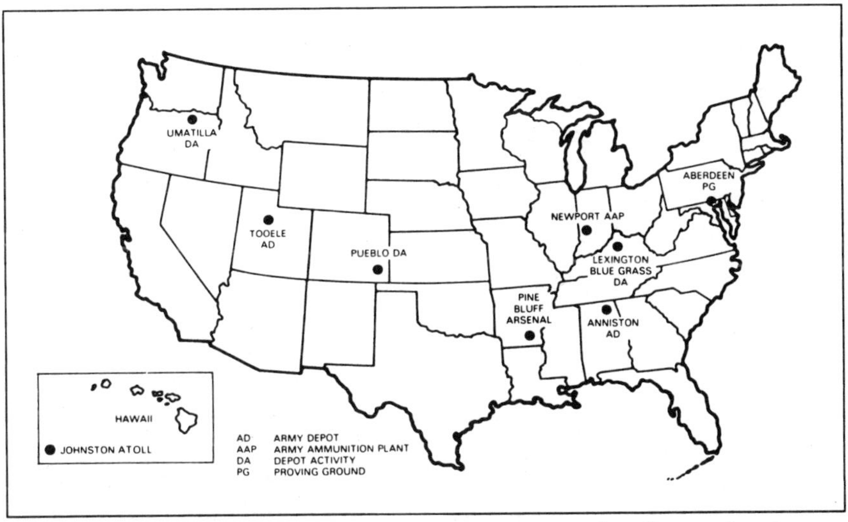

Figure 9-3: Locations of the nine U.S. storage sites for chemical weapons. Adapted from Ref. 15.

Lexington, Kentucky; Newport, Indiana; Tooele, Utah; Aberdeen, Maryland; Umatilla, Oregon; Anniston, Alabama; Pine Bluff, Arkansas; and Pueblo, Colorado (Fig. 3).[15] All munitions and most bulk containers are stored in igloos, which are dirt-covered, reinforced concrete bunkers approximately 90 feet long by 25 feet wide by 15 feet high. An exception is bulk mustard, which is usually stored in the open in 1-ton containers. Also, at Newport 1-ton containers of VX are stored in a building. The manner in which projectiles are packed into a storage igloo is shown in detail in Figure 4.

Environmental Dangers Posed by the Chemical Stockpiles

Except for the new binary weapons manufactured during the past year, the chemical munitions are old and deteriorating. The most serious problems are associated with the M55 rockets.[15] These are loaded with nerve gases, either VX or GB, and have fuses, burster charges, and propellants in place. These rockets are the source of the greatest concern, as they develop the most frequent leaks and pose the greatest threat to civilian populations. Furthermore, although for most munitions the explosion of one device does not cause the sympathetic explosion of others, the M55 rocket is an exception. For the M55 rocket, the detonation of one rocket in a storage igloo would spread to other rockets in the same igloo.[16] There are approximately 2,000 M55 rockets stored in a single igloo about the size of a four-bedroom house (about 2,250 square feet).

In tests conducted at the Black Hills Army Depot and at the Dugway Proving Ground, such propagation did occur.[17] Although 97 percent of

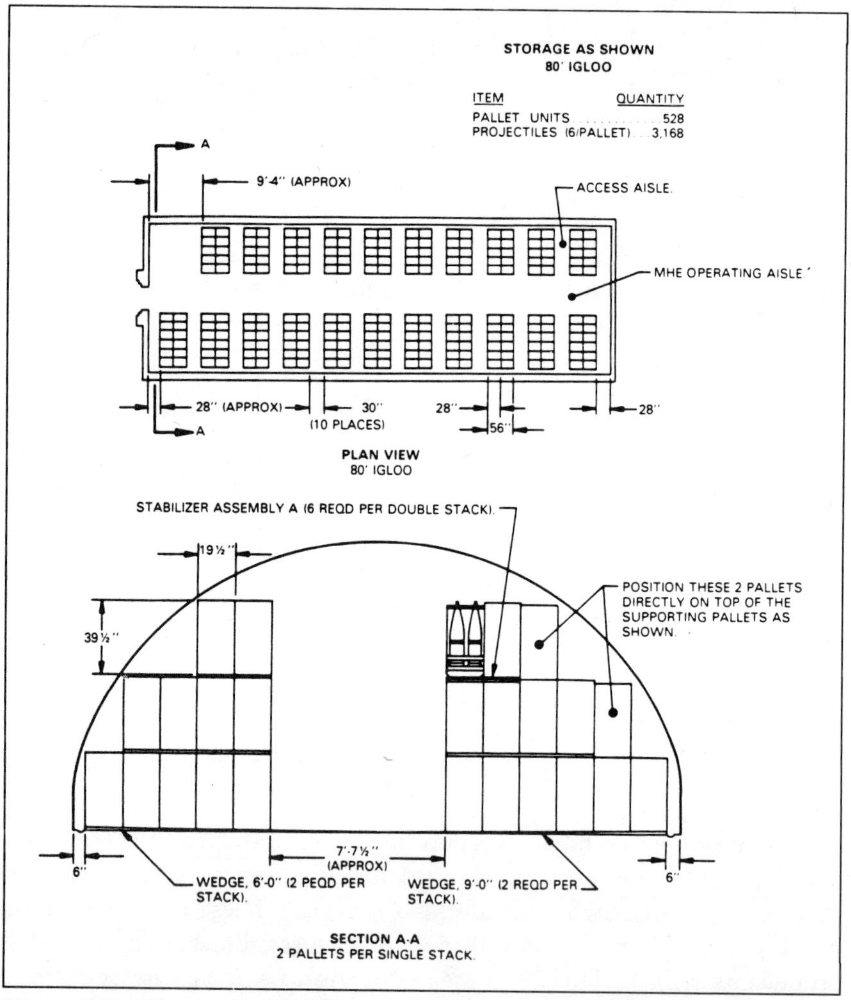

Figure 9-4: Manner in which 3,168 projectiles are stored in one igloo. From Ref. 15.

the agent in the exploding munitions was destroyed by high temperatures in the resulting fire, the Army estimated that 570 pounds of GB were released. For VX-filled rockets, an estimated 37 pounds were released. Most of the nerve gas was released during the first 15 to 20 minutes following the initial explosion, although the munitions in the igloo continued to burn and explode for several hours.

Based on these test results, the Army has estimated the "Maximum Credible Event" for each of its depots. One percent lethality distances were calculated to extend to between 4.5 and 27 miles for explosion of an igloo containing M55 rockets, the value depending on vertical stability of the atmosphere.[18] These calculations are very crude in that they do not take into account the details of the terrain or the variability of atmospheric meteorology. They also assume that only 2.52 percent of the nerve gas would be released, a figure based on a single experiment. Dispersion would approximately follow an inverse square law: thus if 50 percent of the agent were released instead, the 1 percent lethality distances could easily be 4 or 5 times greater, or about 20 to 140 miles from the depot. Emission of chemical species from fires are known to vary over more than an order of magnitude, depending on fire conditions.[19] Also, the simple Gaussian dispersion models used by the Army can only predict what will happen on the average. Plumes of smoke from fires can be transported hundreds of miles before descending on populated areas.[20] The people of Denver, Colorado have been well aware of this fact since the September 1988 fire in Yellowstone National Park.

Clearly, a population of several million currently is at risk from the 40,000 tons of chemical agents now in storage. For example, the Tooele Army Depot, where 42 percent of these toxic chemicals are now in storage, is located only 30 miles southwest of Salt Lake City; the Pine Bluff Arsenal is located 30 miles southwest of Little Rock; the Lexington Blue Grass Depot is 50 miles south of Lexington, but only 3 miles from the city of Richmond; the Edgewood area of the Aberdeen Proving Ground Arsenal is 15 miles northeast of Baltimore; the Anniston Army Depot is 50 miles east of Birmingham, Alabama, and 110 miles west of Atlanta, Georgia; the pattern is clear. Hundreds of thousands of other people live much closer in smaller communities and rural areas.

Release of chemical agents could come not only through leakage or accidental detonation, but also from airplane crashes into igloos, natural disasters such as earthquakes, and terrorist action.[15]

Disposal of Chemical Munitions

Regardless of the outcome of the Geneva negotiations on a chemical weapons treaty, the current chemical weapons stockpile of the United States is considered to be obsolete, and the U.S. Congress has charged the Army to destroy all chemical weapons manufactured prior to 1987. The cost probably will be about $2 billion.[21]

Prior to the Marine Protection, Research and Sanctuaries Act of 1972, most obsolete chemical agents and munitions were disposed of by dumping them into the deep ocean.[15] During 1967 and 1968, for example, 1,706 concrete "coffins," each containing 30 M55 rockets filled with the nerve gas GB, were sunk in the Atlantic Ocean east of the Naval Ammunition Depot in Earle, New Jersey. Other disposal methods have included land burial and open-pit burning.[15]

Improper disposal of toxic wastes associated with production of chemical weapons has created serious environmental problems that are still being dealt with. Throughout the 1950s toxic waste associated with production of nerve gases and insecticides at the Rocky Mountain Arsenal (then 10 miles northeast of Denver) were dumped into earthen reservoirs.[22] The contaminated ground water affected crop yields and killed livestock on nearby farms that made use of water pumped from wells. Furthermore, an estimated 2,000 ducks and other wildfowl were killed annually as a direct result of landing on the arsenal's sewerage reservoirs. As a consequence of lawsuits and political pressure from state officials, the Army decided to drill a 2.5-mile-deep well for disposal of its toxic waste. Between 1962 and 1966, the nearly 165 million gallons of highly toxic waste were pumped into the well—enough to fill a 50-acre lake to an average depth of 10 feet. One month after pumping operations began, the Denver area was hit by its first earthquake in 80 years, and more than 1,500 earthquakes were recorded over the next five years. Whether the earthquakes were caused by the lubrication of a previously inactive fault remains in dispute. In any case, the Rocky Mountain Arsenal is one of the most serious toxic waste sites in the country.

Since the 1972 prohibition against ocean disposal, a variety of options have been considered by the Army, including methods as radical as placing the munitions in an underground cavity and destroying them with a nuclear explosion. The method chosen by the Army and endorsed by a 1984 study of the National Research Council is incineration.[15] To the Army's credit, this method has been used successfully in the past; more than 3,000 tons of mustard gas and 21,000 cluster bombs containing more than 2,000 tons of GB were incinerated at the Rocky Mountain Arsenal in the 1970s, and an experimental pilot plant at Tooele, Utah has been used to burn more than 14 tons of GB and three tons of VX. There were no reported chemical-related injuries or environmental releases.

Because of dangers associated with transporting these munitions to a central disposal site—including the possibility of terrorist attack—the Army plans to incinerate the chemicals at the eight sites where they are currently stored.[22] This poses new dangers to communities located near these depots, and considerable dissension is expected.

The forward-deployed weapons in West Germany pose an international problem. President Reagan stated that the United States will remove its chemical stocks from Army bases in West Germany in 1992.[22] However, transportation across Germany and other European nations is likely to be controversial. A similar problem was encountered in the late 1960s and early 1970s when the Army removed chemical weapons from Okinawa. These weapons were to be transported and stored at the West Coast Army Depot. The uproar was so great that Congress passed a law prohibiting their importation. As a result, the weapons were sent to Johnston Island, about 800 miles southwest of Hawaii, for storage. An incinerator is now being built on Johnston Island, and if all goes as scheduled, these chemical weapons will be the first of the current arsenal to be destroyed beginning in July 1991. It is likely that the German stockpile will be sent there for disposal as well.

Modernization of the U.S. Chemical Arsenal

Following an 18-year moratorium on production of chemical weapons, on December 16, 1987, at the Pine Bluff Arsenal, the U.S. Army once again began filling artillery shells with chemical agents. By a narrow vote in 1986, Congress gave the go-ahead for the production of this new generation of "binary" chemical weapons. Such weapons contain two relatively nontoxic chemical compounds, which are mixed in flight to produce nerve gas. Current projects include artillery projectiles and the Bigeye bomb.[23,24] Binary battlefield rockets and cruise missiles are being considered for production in the late 1990s.[25]

In the artillery projectiles the g-force and rotation of the shell causes the methylphosphonyl difluoride (DF) in one canister to mix with a solution of isopropyl alcohol and isopropyl amine in another canister. The rapid chemical reaction produces GB and about 25 percent of another toxic gas, hydrogen fluoride.[24]

In the 500-pound Bigeye bomb, a core of sulfur is separated by a steel diaphragm from a surrounding liquid: ethyl 2-diisopropylaminoethyl methylphosphonite (QL). A gas generator is started in the bomb, which breaks the steel diaphragm, allowing the sulfur to mix with the liquid to produce VX and some of its isomers. Once the reaction is sufficiently complete, a time-delay fuse ignites to cut the ports in the bomb casing, allowing air to enter and force the liquid to spray out as the bomb falls.[24]

BIOLOGICAL WEAPONS

Biological warfare has been aptly described as public health in reverse. The agents of death include living microorganisms such as bacteria, fungi, rickettsia, protozoa, and viruses, in addition to toxic chemical products (toxins) produced by organisms. The characteristics most dreaded by medical doctors—e.g., toxicity, contagiousness, resistance to antibiotics, resistance to natural immunity, difficulty of diagnosis—are the very characteristics sought for in the biological warfare laboratory. Some of the organisms most actively investigated in these "inverted hospitals" are described in Table 2.

Biological Warfare in World War II

The development of biological weapons was heavily pursued by Japan, the United Kingdom, and the United States during World War II.[2] In the United States, research, development, and testing were carried out chiefly at Fort Detrick, Maryland, where the diseases anthrax, typhus, yellow fever, plague, botulism, brucellosis, glanders, tularemia, dysentery, and many others were investigated for their potential as weapons.[26-28] In addition, large-scale antianimal and anticrop research was conducted there. The first biological weapon manufactured was a four-pound anthrax bomb in 1943. Prototypes were tested at the Dugway Proving Ground near Salt Lake City, Utah. A variety of bombs, aerial sprays, and antipersonnel mines were subsequently developed.

Disease	**Organism**	**Infectivity**	**Effects**
BACTERIAL DISEASES			
anthrax	*Bacillus anthracis*	20,000 organisms of >2 μ inhaled	respiratory form normally fatal if untreated
brucellosis	*Brucella melitensis*	high, 1300 organisms?	long-lasting, recurrent severe fever; rarely fatal
cholera	*Vibrio cholerae*	low by ingestion	severe intestinal infection; sometimes fatal
glanders	*Mallomyces mallei*	high, 3200 organisms inhaled	in acute form, severe fever; often fatal
melioiodosis	*Whitmorella pseudomallei*	high	normally fatal fever producing mania and delirium
plague	*Pasteurella pestis*	high, 3000 organisms	very severe; often fatal
tularemia	*Pasteurella tularensis*	very high, <50 organisms	severe fever; 5-8% fatal
VIRAL DISEASES			
breakbone fever	dengue viruses	high, single mosquito bite, 2 organisms inhaled?	most incapacitating fever known, very rarely fatal
psittacosis		high	mild to severe pneumonia; rarely fatal
smallpox	*Poxvirus variolae*	high, a "few" organisms	severe; often fatal

Table 2: Examples of Pathogens Studied as Potential Biological Warfare Agents

Disease	Organism	Infectivity	Effects
yellow fever		high, single mosquito bite	jaundice-type fever; 30% mortality
RICKETTSIAL DISEASES			
Q fever	*Coxiella burnetti*	very high, 1 organism inhaled?	severe; often fatal
epidemic typhus	*Rickettsia prowazeki*	high	severe; often fatal
FUNGAL DISEASE			
coccidioidomycosis	*Coccidioides immitis*	1350 spores	mild to severe fever; rarely fatal

Adapted from Appendix 2 of ref. 12.

Table 2: Examples of Pathogens Studied as Potential Biological Warfare Agents continued

World War II research into biological weapons by the Japanese made use of at least 3,000 Chinese, Korean, Soviet, American, British, and Australian prisoners of war as human guinea pigs.[28] To test anthrax and gas gangrene bombs, prisoners were tied to stakes with their buttocks exposed to shrapnel from the exploding bombs. The course of the disease was then meticulously tracked and recorded. Other prisoners were infected with cholera and plague and subsequently dissected—sometimes while still alive—to examine the progressive degeneration of their internal organs. Chinese women were infected with syphilis in order to investigate preventive measures against the disease. Details of these experiments fell into the Americans' hands upon the surrender of Japan in exchange for amnesty for the biological warfare researchers.

The Japanese also made extensive use of their biological weapons.[28] It is believed that biological agents were used against at least 11 Chinese cities. For example, in November 1941, Japanese planes dropped grains of wheat and rice and pieces of cotton and paper contaminated with the plague bacterium, *Pasteurella pestis,* over the city of Changteh in Hunan province. The Chinese have estimated that plague attacks alone caused about 700 deaths. It also is estimated that the Japanese poisoned 1,000 Manchurian wells with cholera, typhoid, and dysentery.

Genetic Engineering

The hereditary "blueprint" for all living organisms is DNA (deoxyribonucleic acid), which is contained in the nucleus of every living cell. Through recombinant DNA (rDNA) technology, first conceived of in 1972, it is now possible to remove a gene from one organism and place it in another organism.[29] The rDNA technique is summarized in Figure 5. An enzyme called a restriction endonuclease is used to split the desired gene from the DNA of the donor cell. This donor cell may be plant, animal, or human. The gene could even come from a segment of viral DNA. By use of other enzymes, this gene is then spliced into a plasmid, which is a special type of circular DNA found in bacteria. The plasmid is then inserted into a bacterial cell and the cell is cloned (allowed to replicate to produce many identical bacteria). The gene inserted into the bacterium will code for production of a particular protein.

This rDNA technology has spawned a huge biotechnology industry that so far has focussed on producing pharmaceuticals for the treatment of disease. For example, people with diabetes were formerly treated with insulin obtained from swine or cattle. The cost of producing the insulin was very high. Now human insulin, a protein coded for by a gene contained in the DNA of pancreas cells, is produced by bacteria. Not only is the insulin much less expensive to manufacture, but it is identical to that produced by human cells. All newly diagnosed diabetics in the United States are now treated with rDNA-produced human insulin. The new biotechnology industry that has sprung up to take advantage of this genetic engineering technology holds the promise of making many pharmaceuticals available on a large scale for the first time. The human growth hormone, used to treat children who fail to grow properly, was formerly available only in small quantities obtained from cadavers. It is now readily available through rDNA technology. A promising cancer treatment agent,

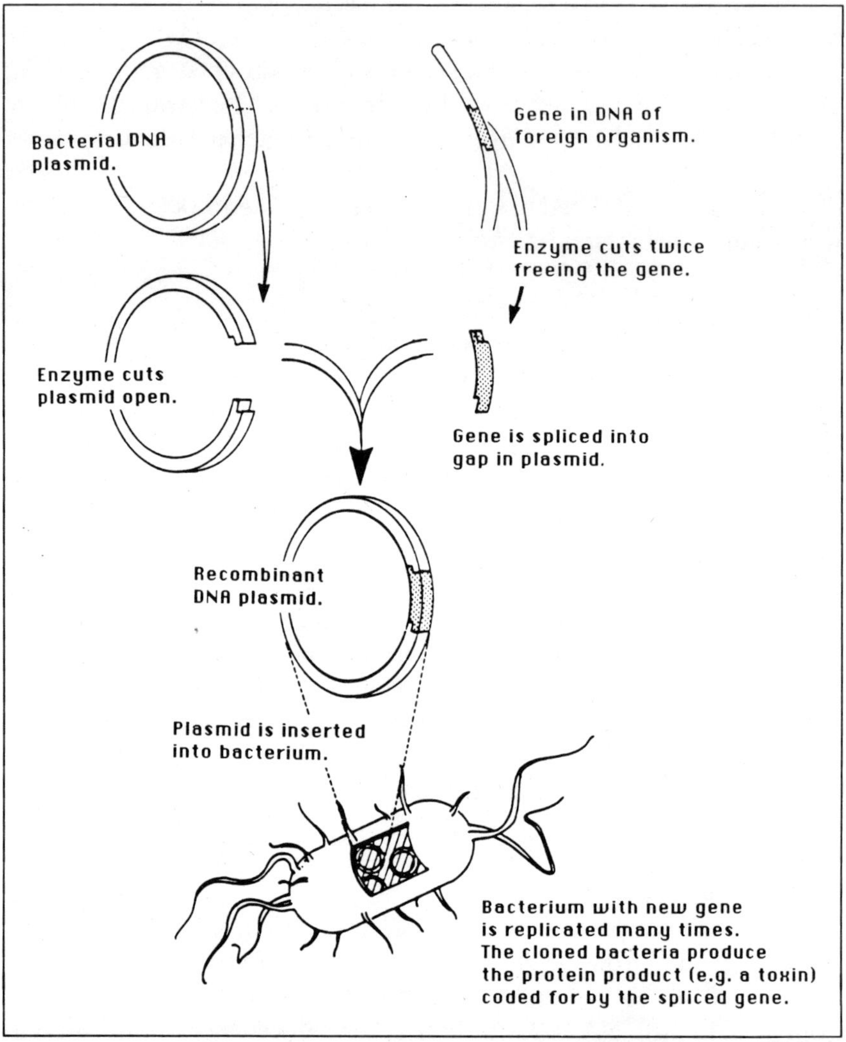

Figure 9-5: Recombinant DNA technique currently used in the biotechnology industry to modify organisms such that they will produce the protein coded for by the "virus" gene. Here, a "virus" gene is a piece of DNA taken from a foreign cell; e.g., the gene for producing insulin obtained from the DNA of a human pancreas cell. In germ warfare applications, the virus gene could come from virtually any organism and code for production of a variety of toxins many times more potent than nerve gas. The "new" bacterium could be used to rapidly produce large quantities of toxins, or it could be used as a biological warfare agent itself to produce a new "designer disease." Adapted from Ref. 28.

interferon, is expected soon to become widely available as a result of genetic engineering.

The newly developed rDNA technology could also be used to produce potent biological warfare agents. Toxins are complex substances produced by animals and plants that can cause the death of other species. Toxins are themselves not alive; they are only the poisons produced by living things. For example, the botulin toxin is produced by the bacteria causing botulism. This along with ricin, a plant toxin, and saxitoxin, a shellfish toxin, were standardized and manufactured in the past by the U.S. military.[28] (These toxins, along with all other stockpiles of biological warfare agents, were ordered destroyed by President Nixon in 1969.) Examples of toxins, their sources, and their relative toxicities are given in Table 3. Just as rDNA technology can be used to produce medically useful proteins that otherwise can only be extracted from large amounts of tissue, rDNA technology can be used to produce toxins that are normally at such low concentrations as to render the extraction impractical. The gene for production of a potent toxin could be taken from a fragile organism, such as coral or jellyfish, and implanted into the DNA of hardy strains of bacteria. Furthermore, hybrid toxins and, in principle, entirely new toxins could now be created and mass produced. On a per weight basis, many known toxins are much more toxic than the nerve gases.

Besides providing an efficient means for producing stockpiles of deadly toxins, genetic engineering may provide the means for making new pathogens with desirable properties for biological warfare such as toxicity (e.g., by incorporating a toxin gene), hardiness, virulence, resistance to antibiotics, and resistance to the natural immune system. Microorganisms can be designed to attack not only humans, but also livestock or specific crops. On the one hand, genetic engineering is used to develop the means of combating disease, while on the other hand, the same technology may be used to invent diseases never known before.

Monoclonal Antibodies

Another relatively new biotechnology, that of monoclonal antibodies (MCAs), is now playing a role in the development of vaccines for use in biological warfare research and development. Animals are injected with an antigen such as a biological warfare pathogen. The antigen stimulates cells in the animal's spleen to begin to produce antibodies against the antigen. These antibodies are complex proteins that bind with specific

Chemical	**Toxin, Source Organism**
	Shiga toxin, *Shigella dysenteriae* (bacteria)
	Tetanus toxin, *Clostridium tetani* (bacteria)
	Botulinum toxin, *Clostridium botulinum* (bacteria) Ricin, castor bean
Dioxin	
	Tetrodotoxin, puffer fish
VX nerve gas	
GD nerve gas	
	Aflatoxin, *Aspergillus,* (fungus)
GA nerve gas	
	T2 toxin, Trichothecene, (fungus)
GB nerve gas	
Mustard gas	Snake venom toxin, rattlesnake
Phosgene	
Chlorine	Bee venom toxin, honeybee

The substances highest on the chart are most lethal, i.e., fatal in the smallest doses. The placement is only qualitative; for example, tetanus toxin is nearly 5 times as toxic by weight as botulinum toxin, and 1.5 million times as toxic as the nerve gas GA.

Adapted from ref. 28, p. 249 and references therein.

Table 3: Relative Toxicity of Selected Chemicals and Naturally Occurring Toxins chemical sites on the surface of the pathogen. However, there are many different binding sites, and different spleen cells produce different antibodies specific to particular sites. By fusing the various spleen cells with tumor cells, it is possible to grow "immortal" colonies of cells in cell culture. Each colony consists of cells producing only one type of antibody. The antibodies may be tested and a colony of cells found that is highly specific to the pathogen. These cells are then further cloned to produce large quantities of the antibody.

The method has led to phenomenal advances in immunization and vaccine purification. Successful implementation of biological warfare would require a highly reliable vaccine for the aggressor. MCAs make possible the identification of specific antigenic sites on the pathogen. Thus, using the rDNA technique, highly specific vaccines could possibly be developed and mass-produced on short time scales.

The Prospect of Ethnic Weapons

Every cell of every human being contains all of the genetic information of that individual. This information is encoded in DNA as a sequence of four different distinct chemical units known as base pairs (adenine, thymine, cytosine, and guanine). It is estimated that this sequence consists of about three billion bases. Most of the genetic information is the same for all people; however, the variations that do exist are what make each person distinctly different from every other person (an exception being identical twins). Various races and populations of people have certain genes (large segments of DNA) in common, and these genes are expressed in a variety of outwardly visible ways, such as skin color. They also are expressed in more subtle biochemical ways, such as susceptibility and resistance to disease.

In a 1970 article in the U.S. Defense Department publication *Military Review*, the Swedish geneticist Carl Larson wrote, "An enzyme deficiency in Southeastern Asian populations [makes] them susceptible to a poison to which Caucasoids are largely adapted. The poison at issue is milk." From here Larson went on to make a case for a diabolical idea that has been around for many years, namely, that genetic differences could provide the basis for developing biological weapons that selectively target certain races or ethnic groups.[30] The possibility of genotype-selective weapons is further explored in the recent book *Gene Wars* by Charles

Pillar and Keith Yamamoto.[31] These authors note that valley fever, an endemic disease of California's San Joaquin Valley caused by the fungus *Coccidioides immitis*, has been studied as a potential biological warfare agent since the 1940s and that blacks are up to ten times as likely to die from the disease as whites. As another example of a natural pathogen having differential effects across racial groups, they cite the Epstein-Barr virus: it produces infectious mononucleosis (a temporarily debilitating but seldom fatal disease) in Caucasians, but for black Africans and Southeast Asians it may cause two different forms of cancer.

Whether or not a sufficiently selective pathogen could be engineered for the purpose of genocidal warfare remains uncertain. Even the mention of the possibility was regarded as "generally irresponsible" by one of the reviewers of this chapter. The idea is, unfortunately, already well established in the open literature. Like the idea of the atomic bomb, it will now always be with us.

The Defensive Loophole for Biowarfare R&D

Signatories of the 1975 Convention on Biological Weapons agree "never in any circumstances to develop, produce, stockpile or otherwise acquire or retain: (1) Microbial or other biological agents, or toxins, whatever their origin or method of production, *of types and in quantities that have no justification for prophylactic, protective or other peaceful purposes*." The italics are mine and constitute the carefully negotiated loophole that allows nations to continue biowarfare research and development.

As in many other military activities, such as the Strategic Defense Initiative, the distinction between offensive and defensive research has become so blurred that "the best offense is a good defense." For example, in order to have a good defense against biological agents it is necessary to first develop and even standardize such agents, so that defensive measures such as vaccines, impregnable materials, and sensitive methods of detection can be developed. In the process, new biowarfare agents (obvious offensive weapons) are developed, and the defensive measures, particularly vaccines, are necessarily highly specific to one's own agents. As such, they become offensive weapons in that they may be used to protect one's own troops and populace, should the new agent ever be used. Taken together, a vaccine and its corresponding agent constitute a powerful weapon system.

Figure 6 summarizes U.S. spending in recent years on "defensive" biological research, according to the Department of Defense (DOD).[32] These figures could actually be much higher as a result of the so-called "black budget," the classified portion of the defense budget, which is blacked out in published material and for which the DOD is not publicly accountable. In any case, the steep increase in funding for research in biological warfare under the Reagan administration is readily apparent.

Another source of concern and further indication of the expanding interest in biological warfare is the Army's plans to build a new biowarfare

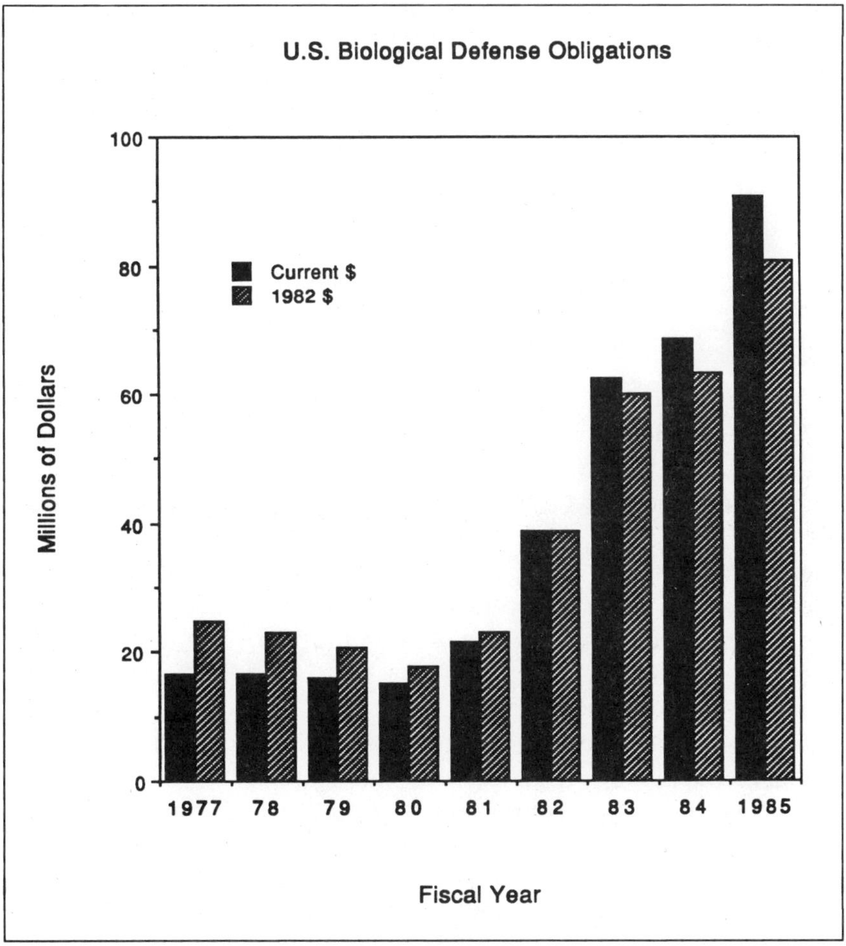

Figure 9-6: Finding of U.S. Biological Defense Research in Dollars and in constant 1982 Dollars. From Ref. 32.

lab at Utah's Dugway Proving Ground.[33] Originally intended to be a maximum containment or Biosafety Level (BL) 4 facility, the lab is intended chiefly to test the degree to which highly infectious agents dispersed in aerosol droplets would penetrate materials and filters that might be used for protection against an attack. Agents originally planned for testing include the organisms that cause tularemia, anthrax, Q fever, and encephalitis.[33] It would also be used to test monitors (undoubtedly based on monoclonal antibodies) for detecting minute amounts of specific biological agents. As a result of fierce opposition from state officials, including Governor Norman Bangerter, the Army has agreed to construct the lab as a lower-level BL3 facility.[34] In fact, the same organisms can still be tested in such a facility but with less security. Testing of bioengineered organisms, however, would require a BL4 lab such as the one in Fort Detrick, Maryland. None of the Army's BL4 labs have provisions for aerosol generation.

Safety in DOD-Sponsored Chemical and Biological Warfare Research Labs

The Senate Subcommittee on Oversight of Government Management (OGM Subcommittee), chaired by Senator Carl Levin (D-Mich.), recently made public its 18-month review of the laboratories conducting research for the Department of Defense in the areas of biological and chemical warfare.[35] The OGM staff found "serious deficiencies in DOD's management of safety issues in the CBW research program, including inadequate regulations, lax safety enforcement, and documented safety lapses." The most serious safety deficiencies were found to be associated with the approximately 150 outside contractors. For those carrying out biological warfare research under the Biological Defense Program (BDP), the OGM staff could find no safety program or safety officers to carry out inspections. No safety reports of the contractors were required by DOD, and it was found that there is no requirement for a contractor to certify decontamination of facilities once the contractor leaves the BDP program. Research involving chemical weapons was found to be monitored by a Chemical Surety Program, but this safety program was also found to have major shortcomings.

Comparison of Biological and Nuclear Weapons

Biological weapons have been and probably always will be unpredictable. In fact, the success in obtaining a treaty outlawing these weapons may be largely attributed to their perceived lack of military utility. Such weapons make little sense for powerful nations like the United States and the Soviet Union, which already possess thousands of nuclear warheads. Biological weapons could be seen by leaders of Third World countries, however, as cheap alternatives to nuclear weapons. A qualitative comparison between biological and nuclear weapons is made in Table 4.

Two factors that have acted to limit the spread of nuclear weapons throughout the world are the scarcity of the raw material uranium and the high cost of processing it into fission-grade material. The gaseous diffusion plant necessary to enrich the uranium in fissionable uranium-235 or the nuclear reactor required to convert uranium-238 into fissionable plutonium-239 carries a price tag in the billions of dollars. Much of the proliferation that has occurred resulted from the export of commercial nuclear reactors.

Biological weapons could be developed at about 1 percent of what it would cost to develop nuclear weapons. The raw materials—microorganisms—are abundant. The facilities needed to work with them may be found in many hospitals or university laboratories. In any case, the

CONSIDERATION	NUCLEAR	VS.	BIOLOGICAL
Cost:	Expensive	vs.	Cheap
Raw Materials:	Rare	vs.	Abundant
Concealment:	Difficult	vs.	Easy
Delivery:	Missiles	vs.	Test Tubes
Time to Arm:	Long	vs.	Short
Physical Damage:	Enormous, Nuclear Winter	vs.	None
Selectivity:	Indiscriminate	vs.	High with Vaccines

Table 4: Comparison of Nuclear and Biological Weapons

required facilities are very inexpensive. For example, the new BL3 laboratory described above that the Army plans to build at Dugway Proving Ground will cost only $5.3 million. Such facilities would be easy to conceal. The biotechnology necessary to engineer new organisms is practiced in many university laboratories where many foreign graduate students are trained, and the techniques are published in the open scientific literature.

The delivery systems for nuclear weapons (ICBMs, cruise missiles, long-range bombers) are as sophisticated and expensive as the nuclear weapons themselves, but a Third World country—or anyone else—intending to use a biological warfare agent would need no expensive delivery system. If the organism is contagious enough, a vial of the material could be carried by a single infiltrator.

Nuclear weapons destroy all physical structures within vast areas. When detonated near the ground, they produce radioactive fallout that can cause casualties among one's own troops and populace. Large-scale use of nuclear weapons would modify the earth's climate, resulting in a "nuclear winter."[36] Biological weapons cause no structural damage, and theoretically, if the population has been vaccinated against the specific agent used, the aggressor could escape with few casualties. However, to be certain that the vaccine actually works would require extensive testing. This caveat is perhaps the best deterrent to the actual use of biological weapons.

The "time to arm" factor is relevant to superpowers as well, and is what makes the "defensive" research and development described above so worrisome. Microorganisms replicate rapidly and thus may be taken from the R&D stage to a stockpiled weapon in a very short time. The production facility might resemble a brewery. This again is in contrast to nuclear weapons, which require many years to enter production.

A NOTE OF CAUTIOUS OPTIMISM

In a few years' time all of the aging and dangerous stores of chemical weapons in the United States will have been destroyed. The new binary weapons, repugnant as they are, are much safer to store and transport than the old weapons. Hopefully, they will be used only to provide leverage for an international treaty on chemical weapons that will require their destruction as well. Progress in Geneva has been slow but steady.[3]

Although biological warfare has been waged on a small scale for centuries, it remains a highly uncertain proposition. The 1975 Convention on Biological Weapons has been signed by more than 100 nations. There have been no reports to date of new organisms designed specifically for biowarfare. However, considering the rapid pace at which molecular biology is advancing, we must institute more stringent international safeguards against biological warfare research and development, particularly as regards genetic engineering possibilities. Such concerns were voiced at a recent review in Geneva of the 1975 Biological Weapons Convention.[37]

NOTES

1. V.J. Derbes, "De Mussis and the Great Plague of 1348," *J. Am. Med. Assoc.* (1966) 196, 59-62.
2. A.H. Westing, "The Threat of Biological Warfare," *BioScience* (1985) 35, 627-633.
3. L.R. Ember, "Fashioning a Global Chemical Weapons Treaty," *Chem. & Eng. News* (1988) March 28, 7-17.
4. *Arms Control and Disarmament Agreements*, Washington, D.C.: United States Arms Control and Disarmament Agency, 1980, pp. 9–18, 120–31.
5. J. Perry Robinson, "Chemical Weapons," In *CBW, Chemical and Biological Warfare*, S. Rose, ed., Boston: Beacon Press, 1968.
6. L.F. Haber, *The Poisonous Cloud, Chemical Warfare in the First World War*, Clarendon Press, Oxford, 1986.
7. *Medical Manual of Chemical Warfare*, 1st U.S. Ed., Chemical Publishing Co., Brooklyn, 1941.
8. J.W. Hill, *Chemistry for Changing Times*, 5th ed., New York: Macmillan, 1988, pp. 569, 576–77.
9. F.J. Brown, *Chemical Warfare, A Study in Restraints*, Princeton: Princeton University Press, 1968.
10. S. Murphy, A. Hay, and S. Rose, *No Fire, No Thunder*, New York: Monthly Review Press, 1984, pp. 111–12.

11. *Chemical Disarmament: New Weapons for Old*, Stockholm International Peace Research Institute Monograph, New York: Humanities Press, 1975, p. 75.

12. R. Clarke, *The Silent Weapons*, New York: David McKay Co., 1968, pp. 53–58.

13. C.R.B. Joyce, "Psychedelics," In *CBW, Chemical and Biological Warfare*, S. Rose, ed., Boston: Beacon Press, 1968.

14. R. Harris, and J. Paxman, *A Higher Form of Killing*, New York: Hill and Wang, 1982, pp. 206–10.

15. *Disposal of Chemical Munitions and Agents*, Committee on Demilitarizing Chemical Munitions and Agents, National Research Council, Washington D.C.: National Academy Press, 1984.

16. *Handbook for Chemical Hazard Prediction*, U.S. Army, 1980. This document states that: "In most planned munitions handling operations, it is not considered credible that explosion of more than one munition would occur in a single accident. Tests of several chemical munitions have shown that sympathetic detonation of the explosives does not occur in normal storage configurations for most items. The two exceptions are the M55 rocket and the M23 land mine."

17. Solomon Irving et al., "Methods of Estimating Hazard Distances for Accidents Involving Chemical Agents," Operations Research Group Report No. 40, Edgewood Arsenal, Md., February, 1970.

18. J.D. Lloyd, Memo from Director, U.S. Army DARCOM Field Safety Activity, Charleston, Indiana, to LTC Leideritz, "Maximum Credible Events (MCE's) and 1% Lethality Distances for DARCOM Chemical Agent Storage Sites," March 2, 1984.

19. A review of the emission factors for a variety of chemical species formed in fires is contained in: "The Effects on the Atmosphere of a Major Nuclear Exchange," Committee on the Atmospheric Effects of Nuclear Explosions, National Research Council, Washington D.C.: National Academy Press, 1985, pp. 108–12.

20. See for example: C.D. Smith, Jr., "The Widespread Smoke Layer from the Canadian Forest Fires During Late September 1950," *Mon. Weather Rev.* 78, 1950, 180–84.

21. L. Ember, "Army Proposes Burning Existing Chemical Weapons On-Site," *Chem. & Eng. News*, July 14, 1986, pp. 24-25.

22. S.M. Hersh, *Chemical and Biological Warfare, America's Hidden Arsenal,* New York: Bobbs-Merrill, 1968, pp. 104–12.
23. J. Perry Robinson, "Binary Nerve-Gas Weapons," In *Chemical Disarmament, New Weapons for Old,* Stockholm International Peace Research Institute Monograph, New York: Humanities Press, 1975.
24. E.M. Spiers, *Chemical Warfare,* Urbana: University of Illinois Press, 1986, pp. 167–74.
25. R.J. Smith, "The Army Is Making 'Repugnant' Weapons Again," *Washington Post* National Weekly Edition, Jan. 3, 1988, p. 34.
26. S.M. Hersh, *Chemical and Biological Warfare, America's Hidden Arsenal,* Indianapolis: Bobbs Merrill, 1968, pp. 68-101.
27. J. Cookson, and J. Nottingham, *A Survey of Chemical and Biological Warfare,* New York: Monthly Review Press, 1969, pp. 75-77.
28. C. Pillar and K.R. Yamamoto, *Gene Wars, Military Control over the New Genetic Technologies,* New York: Beech Tree Books, 1988.
29. See for example: S. Krimsky, *Genetic Alchemy,* Cambridge: MIT Press, 1982.
30. Carl Larson, *Military Review,* November 1970, p. 3.
31. C. Pillar and K.R. Yamamoto, *Gene Wars, Military Control over the New Genetic Technologies,* New York: Beech Tree Books, 1988, pp. 99-100.
32. "Annual Reports on Chemical Warfare-Biological Research Program Obligations," FY78–FY86, U.S. Department of Defense. 1982 dollars are calculated using the Gross National Product Price Deflator index, as used by the National Science Foundation to calculate constant dollars for scientific research and development.
33. C. Norman, "Biowarfare Lab Faces Mounting Opposition," *Science* 240, 1988, 135.
34. L. Ember, "Dugway, Iraq Focus of Chemical Arms Concerns," *Chem. & Eng. News,* Sept. 26, 1988, p. 8.
35. "Preliminary Report of the Majority Staff of the Senate Subcommittee on Oversight of Government Management on DOD's Safety Programs for Chemical and Biological Warfare Research," May 11, 1988.
36. J.W. Birks, and A.H. Ehrlich, "If Deterrence Fails: Nuclear Winter and Ultraviolet Spring," this book.
37. "Gene Splicing Dominates Review of Weapons Pact," *Science* 234, 143–45, 1986.

10

Psychological Effects of Readiness for War

F.J. VAN RHEENEN, M.D.

GARY LAPID, M.D.

INTRODUCTION

WHAT CAN MENTAL HEALTH PROFESSIONALS CONTRIBUTE TO A DISCUSSION OF the environmental consequences of preparing for war? One needs only to look at the language used in the literature to find an explanation. Certain simple yet essential words appear over and over—words like *fear, anxiety, helplessness,* and *denial.* These words refer to occurrences within us: in other words, in the internal environment of each of us. The goal of this chapter, therefore, is to look at what happens in our internal environments in response to preparing for war in the nuclear age. Jung said, "The history of mankind hangs on a thin thread, and that thread is the human psyche."

In this chapter, we will consider some of the ways that the preparation for war is affecting the human psyche. We will look at both the negative and positive consequences to our individual internal environments, as well as to the species as a whole. The advent of atomic weapons in 1945

We would like to express our appreciation to Kathy Shade for her research and creative input, to Julie Marx for tireless work as our administrative assistant, and to Eileen Moran Van Rheenen for her editing prowess.

marks one of the most important moments in modern history. It was a transformational moment—a moment resulting in an abrupt shift in the conditions affecting human evolution.

Since 1945 our species has been endangered by its own destructive potential. Global, national, and individual security has a new definition. With the ability to split matter, a new archetypal symbol—the mushroom cloud—has been inserted into the collective unconscious of the species. This *cumulofungus* archetype is so overwhelming as a symbol that it is difficult for the psyche to digest it. The presence of this lasting image has altered our perceptions and profoundly changed our psychological milieu.

Three of the most obvious psychological alterations are a decreased sense of control, an increased sense of anxiety, and an enhanced experience of conflict. Perhaps the most significant is the diminution of our illusion of being in control. The end result of using these kinds of weapons cannot at this point of technological development be controlled. Thus these weapons threaten the user as well as the targeted recipient.

A decrease in our sense of control leads to an increase in anxiety. Anxiety in turn results in several characteristic human responses.

- One response is to try desperately to regain control on a nationwide scale. The strategy for so doing adopted by conservative policy makers has been to increase the number of weapons produced and maintained. This group favors a technological solution as a response to the anxiety. Such persons are in search of weapons that would ultimately give their country the edge, thereby reinstating the kind of security and control experienced prior to the nuclear age.
- A second kind of response—also on a nationwide scale—is to try to get rid of the weapons or to reduce the number of weapons significantly. Those who respond in this way believe that the more weapons with mass destructive capability the nation has, the less secure it is. This group points out that every time one superpower escalates the arms race, the other superpower responds in kind, producing an ever-increasing number of deadly weapons on our planet.
- A third response is denial or numbing. Seeking immediate relief from the anxiety, this group uses primitive ego defense mechanisms that enable them to disregard the entire issue and to continue business as usual. Individuals in this group usually want to leave decisions on national security to the experts.

- A fourth response to loss of control is the effort to regain control on an individual basis by attempting to save one's own skin. This leads to elaborate preparations for nuclear disaster, such as the building of personal shelters, the stockpiling of food and weapons, and other survivalistic maneuvers.
- A fifth response is prayer and the appeal to a higher authority. This means rejecting personal effectiveness and putting the outcome into the hands of God.

Loss of control leads to the sense of impotence and a decreased sense of efficacy. Again, there are a few common human responses to this. One is to "give up and give in"—to feel overwhelmed, defeated, and increasingly passive. This leads to more and more citizens' disenfranchising themselves, leaving vital decisions to a smaller and smaller power elite. Many individuals relinquish their personal vision for the future; they become depressed or seek immediate gratification via sex, drugs, or other forms of escapism. For others, loss of efficacy results in rage—impotent rage. This rage can be acted out destructively, as in crimes against property or persons, and increased selfishness and greed. Alternatively, the rage can be acted out constructively through enhanced involvement in the political process, willful resistance, or considered action that demonstrates one's convictions.

The presence of weapons of mass destruction does not simplify any aspect of human existence. In fact nuclear weapons enhance and intensify intrapsychic, interpersonal, and international conflicts as well as conflict with nature itself.

There are many examples of enhanced intrapsychic conflict resulting from the presence of weapons of mass destruction:

- The battle between the part of us that wants to compete, prevail, and win, versus the part of us that longs to cooperate, coexist, and merely survive;
- The battle between the curious, scientifically probing part of ourselves and the part that fears that hubris will lead to punishment for getting too big, for meddling too much with nature;
- The battle between the part of us that fears the expansion of and domination by authoritarian governments and the part of us that fears confrontation in a nuclear holocaust;
- The battle between the deep, often unconscious part of us that wants to get off the treadmill—i.e., wants to die—and the part of us that cherishes life and looks ahead with enthusiasm.

The presence of nuclear weapons heightens interpersonal conflict in several ways. The question of whether to continue an expensive and dangerous arms race or to disarm and face potential domination by our enemies has divided families, friends, churches, and other institutions. Because so much is at stake, Americans—and many others—are divided vehemently on this issue. In terms of our military policy, "hawks" and "doves" are fighting over the helm. Each side would like to determine the direction of our country's policy. This struggle creates the potential for a contradictory and unwieldy foreign policy just when we are most in need of clear direction. There is disagreement about how much of our tax dollar should be spent on military security and how much should go to support social programs. There is the further question of the patriotic obligation to support the foreign and military policies of our elected officials, versus the obligation to speak out if those policies violate one's sense of propriety, ethics, and good judgment. Finally, there is disagreement about whether to focus energy on short-term goals and strategies or to approach the nuclear issue with more global long-term solutions.

In the international domain, the development of weapons of mass destruction has changed the complexion of geopolitical conflict. Perhaps the development of nuclear weapons has prohibited all-out global war, as some would argue. On the other hand, regional conflicts could now easily escalate to a nuclear holocaust, and therefore minor border skirmishes are riskier than ever. Nuclear nations posture and joust with each other and pay lip service to the need for nuclear ethics while surreptitiously trying to get the upper hand. Nonnuclear nations often resent being held hostage by the nuclear nations. Great confusion abounds, as manifested by the recent bitter battle in Western Europe about whether it is safer to have intermediate range missiles on one's soil or not. The world has become more polarized, with the superpowers establishing large spheres of influence. There is intense competition to enlist nonaligned nations, and geopolitical power vacuums attract both sides. There is also the conflict between nuclear nations that are willing to supply nuclear weapons, fissionable materials, and scientific know-how to nonnuclear or emerging nations, versus those nations that resist and prohibit this lucrative practice.

Conflict with nature is also intensified by the presence of nuclear weapons. It has been estimated that the detonation of a very small fraction of the world's nuclear arsenal could result in a nuclear winter (see chapter 7). Even a very limited nuclear exchange could result in massive fires. Increasingly, human beings seem to be at war with nature, treating the living Earth as an inanimate object. Underground testing rattles the foundations of our homes. Radioactive materials from waste sites, accidents, seepage, and test fallout poison our land, our water, and the air we breathe. We have been too cavalier with our planet, and many scientists fear future generations will suffer the consequences. Native American tribes used to ask what effect a proposed action would have on the next seven generations. Modern humanity might return to this intelligent practice.

We will now focus on four areas of particular importance to the psychological aspects of preparing for war. These are: denial, helplessness, belief systems and the image of the enemy, and effects on children.

DENIAL

One of the most frequently used words in the literature on the threat of nuclear war is denial. Denial is a defense mechanism used to avoid awareness of painful emotions—a way to deal with an unwelcome thought, impulse, or feeling by rejecting its significance or importance. It can be a normal process used to help one accept bad news. Elizabeth Kubler-Ross[1] describes denial as a common first reaction when a person is told of the diagnosis of cancer. It is maladaptive if it interferes with rational action such as seeking proper medical care.

Robert Jay Lifton[2] writes about a form of denial specific to the threat of nuclear war. He calls it psychic numbing. This is a condition in which we live our lives on two planes: knowing that we can be annihilated in a moment, while carrying on business as usual.

Mental health professionals worry that psychic numbing about the nuclear war issue may lead to psychic numbing about many issues. When people shut down their feelings to avoid pain, they can rarely stay within that focus. Shutting down feelings in order to diminish the experience of pain usually means diminishing the experience of pleasure as well.

Denial is a difficult defense mechanism to assess. We have no blood test, X-ray, or CAT scan for denial. In spite of the difficulty of measuring it, we believe denial of the nuclear threat is widely present in adults as well as in young people. One of the most convincing arguments for the significant presence of this denial is found in the Goldenring and Doctor[3] study of worry about nuclear war among 900 California teenagers. When the researchers asked the youngsters to name three things about which they worried the most, only 12 percent listed nuclear war or war in general.

However, when students were given a list of 20 items to rank and nuclear war was included, nuclear war moved up to the third biggest worry, preceded only by the death of parents and getting bad grades. In the last part of the test, the students were asked to go back over the list of 20 items and choose their top five worries. In this instance, nuclear war was second, outranked only by parental death. Most striking is the finding that the two most powerful worries—parental death and nuclear war—were hardly mentioned when open-ended questions were asked. Only when they were offered as choices among 20 possible worries did they become the first and second major worries by a wide margin. This suggests these are denied fears that can be brought to consciousness when directly confronted.

Since there is no defense against nuclear weapons, perhaps the best defense is against feelings about nuclear weapons: it is psychologically safest to distance ourselves from the prospect of nuclear war and the fear of that war.[4] Yet, as painful as accurate perceptions are, they are healthier and offer the potential for more genuine well-being than denial. Accurate perceptions can also motivate involvement and optimism. Many studies[5-10] have shown that concerned students (those with less denial) felt optimistic that their involvement could help prevent nuclear war. Those experiencing painful perceptions are often the ones who will work for peace and disarmament. Jung put it succinctly: "There is no birth of consciousness without pain."

HELPLESSNESS

Denial is probably the most common and consistent reaction to anxiety produced by the nuclear threat; however, another very common response is a feeling of powerlessness or helplessness. Helplessness translates into the sense of having no control over one's destiny. Thus individuals feel ineffectual, hopeless, and fatalistic, believing that "If it happens, it happens, but there is nothing I can do about it." Such people put their faith in the government and the "experts," having no faith in themselves as effective in the prevention of war. For obvious reasons, children and adolescents are particularly susceptible to feelings of helplessness. They seldom see themselves as having any real power or efficacy in determining policy in the adult world.

Earlier we suggested that there is an inverse relationship between control and anxiety. When one's sense of control is reduced, one's level of anxiety goes up, and an escalating level of anxiety is motivating, up to a certain point.

In medicine there are many examples of relationships demonstrable in an inverted-U graph. Simply stated, a certain amount of something is productive but too much is paralyzing. Examples include the effect of heart rate on the amount of blood the heart is able to pump, the effect of stimulants (caffeine and amphetamines) on work output, and, of particular relevance to this discussion, the effect of anxiety on performance. In Figure 10-1, the graph indicates quality of performance on the Y-axis and anxiety on the X-axis. As anxiety increases, it stimulates enhanced performance, up to a point. Past this point, the level of anxiety is debilitating. Being overwhelmed by anxiety is one of the major causes of the experience of helplessness.

BELIEF SYSTEMS AND IMAGE OF THE ENEMY

One should never underestimate the power of belief. Beliefs can greatly cloud perception and distort reality. For example, many individuals lack self-esteem as a result of destructive experiences early in their lives. These individuals are likely to think of themselves as unworthy, unintelligent, or

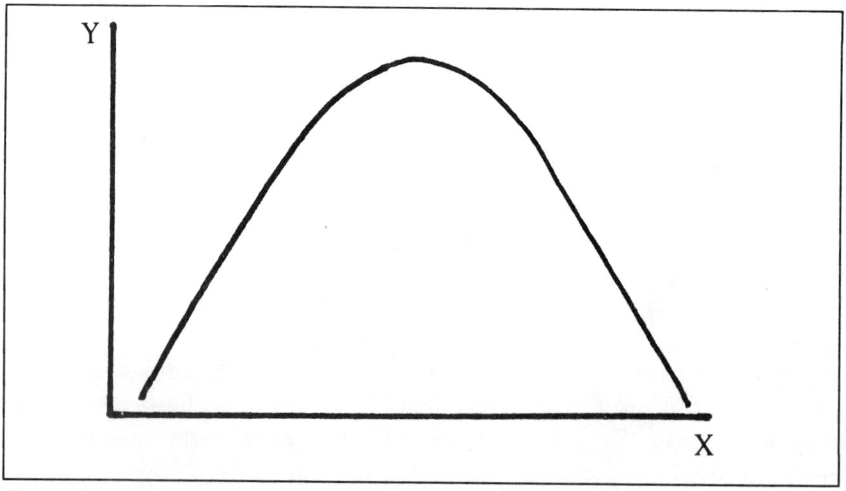

Figure 10-1: Graph showing relationship of anxiety levels to the quality of performance in individuals

unattractive in spite of convincing evidence to the contrary. Faulty or neurotic beliefs beget selective perceptions. Thus, individuals with low self-esteem will hear and accept any criticism since it confirms their low opinion of themselves. Likewise, they will selectively ignore or dismiss praise and compliments since this contradicts their negative self-image and beliefs. Cognitive dissonance is the dynamic by which we humans selectively ignore or reject data that threaten current beliefs, and embrace data that reinforce current beliefs.

The lack of concordance between reality and perception due to the strength of one's belief systems occurs not only on an interpersonal level but also on an international scale. Nearly every nation promotes notions of its allies and its enemies. Human beings have a natural tendency to simplify by polarizing the world: "alike vs. unlike," "good vs. bad," "me vs. not me." This method of classification greatly simplifies our judgments about varied happenings of an immensely complex world. In addition, it may help make distant events seem personally relevant and meaningful. Individuals who perceive that their beliefs and actions are similar tend to form groups. As the group grows, the ideas and beliefs that the group supports become more well defined and often binding on participants. As the beliefs and characteristics that define the group are solidified, the affiliative bonds between group members are strengthened. Those outside the group become viewed as "not me," unlike, alien, different, or foreign. These perceptions generate fear, distrust, and alienation.

Jerome Frank[4] states that "When two groups compete for the same goal, this distrust rapidly escalates into what has been called the image of the enemy. . . . Enemy images mirror each other—that is, each side attributes the same virtues to itself and the same vices to the enemy." If the members of one group wish to believe that they are "right" and "good," then the enemy, or individuals outside of their group, are consequently "wrong" and "bad."

Images of the enemy develop out of fears of what is different, unknown, or in competition with what a particular group wants and believes. A group supports its image of the enemy by maintaining a belief system that consistently highlights the differences between the two camps. Furthermore, a given group tends to project onto enemy groups negative qualities that the given group defines as undesirable or unacceptable. Carl Jung called this phenomenon "the projection of the shadow." Thus, disowned or denied negative qualities of a group are attributed to the "other." We then feel at peace with ourselves and can justify our hatred of the alien group, which is now encumbered with the negative qualities projected onto them.

Another common psychological dynamic that comes into play between individuals as well as nations is that of attribution. What this means is that I will perceive a certain negative quality in you and attribute it to some flaw in your character or ethics; I will view you as wrong or inferior for possessing or manifesting this negative quality. On the other hand, if I see that same negative quality in myself, I will attribute it not to a character or ethical flaw, but to external circumstances or conditions. Thus, I justify my behavior as necessary, reasonable, or unavoidable given the situation.

These are the kinds of mechanisms that produce and maintain the reciprocal distortions of perception between the United States and the Soviet Union. For example, when the Soviet Union shot down the Korean airliner in August of 1983, Americans embraced this as further proof of the Soviet Union's ruthless, treacherous, and untrustworthy nature. Yet when the United States shot down an Iranian airliner in 1988, it was justified as a defensive act in the midst of a naval battle. Furthermore, most objectively prosocial or humanitarian acts by the Soviet Union have been dismissed by the American press and public as a propagandist ploy. In either case, the belief system acts to maintain itself. Individuals and the media go through a process of screening in which those messages that confirm existing beliefs are accepted and those that oppose existing beliefs are rejected.

If we do not mature beyond these simplistic mental mechanisms, we will be unable to secure our future and prevent crises. One of the most tragic possible outcomes for our world would be destruction based on distorted perception or misrepresented reality. The former ambassador to the Soviet Union, George Kennan, addressed this concern in a speech in November 1981:

> I find the view of the Soviet Union that prevails today in our government and journalistic establishment so extreme, so subjective, so far removed from what any scrutiny of external reality would reveal, that it is not only ineffective but dangerous as a guide to political action. This endless series of distortions and over-simplifications; . . . this reckless application of the double standard to the judgement of Soviet conduct and our own; this failure to recognize the commonality of many of their problems and ours . . . are not the mark of the maturity and realism one expects of the diplomacy of a great power.[11]

Early in his presidency, Ronald Reagan contributed to the stereotyped image of the enemy when he called the Soviet Union "the evil empire." Yet after the May 1988 Summit Conference in Moscow, Reagan retracted this, saying he no longer saw the Soviet Union as "the evil empire" and that the phrase belonged to an earlier time.[12] In his farewell news conference in December 1988, Reagan amplified those conciliatory remarks by saying of the Soviet Union, "I do think there is evidence that they don't like being the pariah, that they might want to join the family of nations."[13]

Other historical examples exist that reveal a change in the American image of the enemy. Jerome Frank cites evidence for this by comparing Gallup polls from 1942 to 1966.

> Back in 1942, Germany and Japan were our bitter enemies, and Russia was our ally; and in 1942, among the first five adjectives chosen to characterize both the Germans and the Japanese were: "warlike," "treacherous," and "cruel." . . . In 1966, Germany and Japan had become our loyal allies, and Russia was more or less our enemy, . . . and the Russians were now "warlike" and "treacherous." In 1966 when Gallup surveyed responses to mainland China, predictably the Chinese were seen as "warlike," "treacherous," and, being Orientals, "sly." After President Nixon's visit to China, however, almost immediately these adjectives disappeared about the Chinese and they are now characterized as "hard-working," "intelligent," "artistic," "progressive," and "practical."[14]

The rapidity of this change of attitude about a whole nation supports the concept of just how subjective and projective such assessments are.

As we move further into the nuclear age, we are beginning to understand that national security is tied to global security. When two nations possess nuclear weapons, the notion of "I win; you lose" is outdated. Given this reality, it is imperative that anachronistic belief systems and mental mechanisms such as those described here be discarded and replaced with beliefs and dynamics that are appropriate to the technology we have created.

IMPACT OF NUCLEAR PREPARATION ON CHILDREN

Judith Vigna's children's book entitled *Nobody Wants a Nuclear War* begins with a child stating, "Sometimes I'm scared there'll be a nuclear war and I'll never grow up."[15] Our concern here is the extent to which children are fearful about nuclear war and how that fear affects their lives.

Some things in life seem obvious. It doesn't take an elaborate research project to know that children are adversely affected by the death of a parent, divorce in their family, the loss of a sibling, or parental substance abuse. It seems common sense that if children fear the death of their parents, it follows that they would be at least as fearful if not more so about the totality of destruction in a nuclear holocaust—losing their parents, their siblings, their friends, their pets, and their lives.

We have reviewed the literature on the psychological effects on children of preparing for nuclear war. Many of the studies can be criticized for their methodological limitations. Some authors[16,17,18] criticize the research for confusion in the definition of terms, loaded questions, unrepresentative samples, and an overuse of self-reported data. They suggest that the righteousness of the cause and the felt need for results may override any effort at scientific rigor. It is certainly possible that professionals who feel strongly about the hazards of nuclear preparedness might look for results that confirm their beliefs. However, it is also possible that the prospect of nuclear war has a significant negative psychological effect on our children.

The most important findings in the literature on this subject can be summarized as follows. Half of the children surveyed had become aware of nuclear war by age 10, and by age 13 almost all children were aware of it. Somewhere between one-third and two-thirds of the youth of the world feel anxious and worried about nuclear war. A significant segment (40-50 percent) of young people say nuclear war is either their biggest worry or second only to parental death. Many young people see nuclear war as likely in their lifetime and feel that this uncertainty about their future produces an immediate effect on their lives.[5-10]

Some studies[5] identified two distinct groups of students: a concerned worried group and a despairing group. Those who worried showed a positive correlation between being informed, interest in government, political activism, and optimism that nuclear war could be prevented. Those who despaired felt that nuclear annihilation was likely, and evidenced dissatisfaction and little desire to control their future. They were poorly informed on nuclear topics and showed little interest in government or political activism.

There are only a few studies in the literature aimed at the attitudes and reactions of Soviet children. The findings of Chivian's 1985 study[9] of 485 Soviet youths were confirmed and expanded by his recent more extensive study.[10] In the latter, Chivian and his colleagues compared over 2,000 Soviet youths with more than 3,000 of their American counterparts. For both the Soviet and American children, the two greatest worries were nuclear war and death of a parent. Ninety-four percent of the Soviet sample indicated they were "very worried" about nuclear war, and 54 percent of American youths were similarly affected. There was agreement between the students in the U.S. and U.S.S.R. about survival following a nuclear war. At least 70 percent of both groups agreed that a "nuclear winter" would follow a worldwide nuclear war and that almost no one would survive. The Soviet youths were considerably more optimistic than their American counterparts that nuclear war could be prevented and that there could be cooperation and friendly relations between the United States and the Soviet Union. Perhaps this may be related to the fact that most Soviet youths are actively involved in peace-related activities in their schools and summer camps.

American children received their nuclear information from television, school, newspapers, and parents—in that order, with parents a distant fourth. Half of the young people said they never discussed the topic with their parents.[3] This finding suggests that for many in America discussion of the nuclear issue is taboo, thus indicating the degree of fear and anxiety associated with it.

In spite of these suggestive studies, Beardslee states that "no study of the impact of nuclear war on children and adolescents has demonstrated any serious psychopathological effects that have resulted from the threat."[19] Young people definitely report they are worried, but it is difficult to assess the impact of that worry on behavior. Many of the young people interviewed express pessimism about the future. They also describe feelings of fear and hopelessness, and have a hedonistic and materialistic focus on the present. It is difficult to separate the nuclear threat from other stressful factors such as economic troubles, declining American prestige, disillusionment with the political system, family breakdown, and drug use.

We have discussed how anxiety, despair, and helplessness affect our children in their reaction to the threat of nuclear war. Nuclear weapons can damage our country without ever being detonated. President Dwight D. Eisenhower said, "The problem with defense is how far you can go without destroying from within that which you are trying to defend from without."

Between 1981 and 1988, $2 trillion, or $743 million per day, was invested by the United States in a massive peacetime arms buildup. At the same time, the world's largest creditor nation became the world's largest debtor nation, with a $2.6 trillion national debt.

When parents decide to spend limited family resources on drugs and alcohol rather than books and food, the mental and physical well-being of their children suffers. Similarly, while attempting to keep ourselves the world's number one military power in an arms race with the Soviets, we have fallen behind in our overall social development: worldwide we rank fifth in literacy rate, seventh in public education expenditures, eighth in life expectancy, sixteenth in percentage of women in higher education, and twentieth in number of children per teacher.

Willingness to protect children, the ill, and the elderly has always been a moral litmus test of any decent and compassionate society.[20] This kind of erosion of the infrastructure of our social institutions tends also to erode our sense of well-being and national pride. More and more Americans are questioning our country's priorities. The nuclear age necessitates revitalized ethics, which we must start groping to attain.

THE SILVER LINING IN THE MUSHROOM CLOUD

There is an old Taoist story about a peasant village in China. The village was poor and therefore known as an unfortunate village. One man in the village had a horse, and could plow more than anyone else; this man was known as the fortunate man in the village. One day the man's horse ran away; then he was known as the unfortunate man. Two days later, the man's horse returned and brought with it a wild horse from the environs. Then the man was again known as the fortunate man in the village. Three days later the man's son tried to ride the wild horse and was thrown off and broke his leg. Then the man was known again as unfortunate. One week later the army came through town to take all the able-bodied young men to war, and they didn't take the man's son because he had a broken leg. Then again the man was known as the fortunate man in the village. *And so on.*

We have all experienced seeming tragedies or mishaps in our lives that turn out to be blessings in disguise. Therefore, we may ask whether the development of nuclear weapons and our current nuclear predicament is fortunate or unfortunate, but it is too soon for an answer. Whether or not the development of nuclear weapons and other weapons of mass destruction is fortunate or unfortunate will be a function of how our species responds to the evolutionary pressure and challenge of the weapons.

The development of nuclear weapons was inevitable. Nuclear fusion warheads are basically small stars. Given humankind's insatiable curiosity and undaunting drive to replicate nature, it was inevitable that someday we would learn to split the atom and then combine that with fusion reactions to create small stars. Throughout human history, there has been a reciprocal relationship between technology and consciousness. When the first primate picked up a bone and used it for a hammer, it changed his way of thinking. This changed cognition can then create new tools that exert additional evolutionary pressures, a process that continues ad infinitum. As we approach the 21st century, we must face the challenge to elevate our consciousness sufficiently to assure that we act responsibly and appropriately to the weapons of mass destruction that we have created.

In terms of Jungian psychology, the nuclear age has produced two new archetypal symbols in our collective unconscious. One is the mushroom cloud, image of mass destruction and collective death. The other is the image of the Earth, as seen from outside the frame of reference of the Earth, as in the pictures of our home taken from space. Both images are products of human ingenuity and creativity. Both images are the result of modern technological expansion. One image seemingly represents evolutionary potential, the other, evolutionary demise. All of us are being influenced by these archetypal symbols. How we as a species will consciously respond to their influence is uncertain at this moment.

We can say that to date the human species has been remarkably adaptive. We have met threats to our survival with courage and creativity. Thus we have continued, we have flourished, and we have evolved. Most major threats and evolutionary pressures, however, have come from outside of ourselves. Now, the major threats to our collective survival are the result of our own doing. Successful adaptation will require some major attitudinal shifts, including the way we relate to each other, the way we solve conflicts, and the way we relate to the Earth.

The development of weapons of mass destruction creates a new existential reality. We now have the ability to destroy everything with a single decision—the push of a button or the turn of a key. Never before has this ability existed. Even if we did manage to eliminate all existing nuclear weapons, the recipe is known and the potential to destroy everything is forever with us.

All existential realities have positive (constructive) and negative (destructive) potentials. For many, the negative or destructive potential of our new existential reality appears more prominent than the positive potential. When we look through the eyes of fear, the development of nuclear weapons is seen as an unfortunate event in human history. If we don't look through the eyes of fear, however, we begin to see the fortunate or positive evolutionary potential in the development of nuclear weapons.

Perhaps the most immediately positive possibility is the potential for humankind to learn to neutralize its aggression and resolve conflicts without resorting to violence. Throughout history, the option of all-out combat has followed the breakdown of diplomacy. With the advent in recent times of weapons of mass destruction that do not honor national borders and cannot be easily controlled or contained, the boundary between homicide and suicide becomes blurred. This creates an imperative to curb our violence and find new means (other than "might is right") to solve conflicts.

Marshall McLuhan and others have suggested that nuclear weapons be displayed in front of major governmental agencies and monuments to remind us over and over again of the imperative to avoid all-out conflict. Many have suggested that the absence of a major world war in the past 40 years is attributable to the availability of weapons of mass destruction. These weapons challenge the species to resolve conflicts differently. To have the power to destroy and not exercise it enhances self-esteem. When the taboo against the use of nuclear weapons and other weapons of mass destruction ultimately becomes primary to all cultures, our species will have succeeded in accomplishing the important developmental step of becoming truly responsible for our destructive capacity.

The increasing awareness of interdependency on each other and on the Earth is a second glimmer of hope. During the last century or so, human beings (as well as nation states) have acted largely as if they were isolated units driven by competition and the wish to dominate and control. We have blindly and urgently used up natural resources and lived as though there were no tomorrow.

Today, as we move into the age of global communication, a post-modern world view is emerging. Television and other communication networks make it possible to be in contact with the entire planet instantaneously and simultaneously. People of different cultures are getting to know each other as never before. We are recognizing our economic interdependence as manifested by the domino effect that occurs around the world when a major economic power begins to falter. We are becoming increasingly aware of our ecological interdependence—if I poison the atmosphere or destroy the ozone it definitely affects you and vice versa.

The presence of nuclear weapons makes us cognizant of our interdependence in yet another way. Nuclear weapons are forcing us to learn the lesson of coexistence even with people we don't like, don't approve of, or don't trust. The value of this increased interdependence is that it discourages domination and encourages cooperation. Shifting our emphasis from mutual assured destruction to mutual assured development is likewise served by true recognition of our interdependence. Finally, the recognition of interdependence helps the shift in modus operandi from "me or you" to "me and you." Paradoxically, by creating forces with the capacity to tear our entire existence asunder, we have created a potential counterforce as well, allowing us to bridge differences in culture and points of view. Linked together by the shared desire to survive, we are beginning to generate a rudimentary harmony out of our unique and varied individual cultural rhythms.

A third ray of hope resulting from the development of weapons of mass destruction is the emergence of a spirited and informed citizenry throughout the world. Individuals and groups from all cultures are rising up and speaking out. National and international networks of responsible and dedicated people are responding to the common danger with vision and the quest for revitalization. More and more human beings, voluntarily transcending self-oriented day-to-day concerns, are seeking creative solutions to the current threats to our collective survival. Perhaps never before has such a force existed internationally.

Related to the emergence of the informed and committed citizenry is the evolution of a species-wide ethic—a postmodern world view that is appropriate to dealing with the current threats to our human survival. This ethic calls for transcending domination politics, greed, short-term gain, and environmental irresponsibility.

Another fortunate outcome of our current existential reality is the opportunity for us to look squarely at our human arrogance and technological hubris. For the last 100 to 150 years, we have been addicted to development and control. As a result, we have not been acting in harmony with our environment. Instead, we have attempted to control and dominate anything that resisted our "progress." Thus, our world is teetering out of balance. The development of weapons of mass destruction is giving us pause, however. With the capacity literally to destroy everything, we have to ask ourselves, What are we doing? Similarly, current environmental imbalances and destruction in the service of progress and development are forcing us to question our judgment. Major mechanical failures and accidents are becoming more common and more visible.

All this is creating an international wave calling for heightened consciousness about technological development. There is a growing withdrawal of blind trust in technology and a questioning of the wisdom of technological solutions to human problems. Likewise there is a growing call for ethical considerations vis-a-vis development. Just because we can produce or create something is not a good enough reason to do so. We must begin to ask hard questions, such as: Will this product (and by-products) enhance or detract from the quality of human existence? Technological hubris is being called into question as a result of many tragic accidents and mishaps such as Chernobyl, Three Mile Island, and the explosion of the space shuttle Challenger.

One final opportunity is the challenge to our species to grow out of its adolescence into full adulthood. Power and responsibility must go hand in hand. Just as an adolescent must raise his or her level of integrity when getting behind the wheel of a car, our species needs to raise its level of integrity now that we have learned how to harness atomic energy.

Roger Walsh in his book *Staying Alive* says it this way:

> We have created a world situation that appears to demand unprecedented psychological and social maturation for our survival. Until now we have been able to cover or compensate for our psychological shortcomings. We have been able to consume without fear of depletion, discard wastes without fear of pollution, bear children without fear of overpopulation, and fight without fear of extinction. We have been able to act out our psychological immaturities rather than having to understand and outgrow them, to indulge our addictions rather than resolve them, and to revolve through the same neurotic patterns rather than evolve out of them. But if all the world is a stage, it is now no longer a big enough one for us to continue playing out our psychological immaturities. It is time for us to grow up and we ourselves have created the situation that may force us to do so.
>
> This growing up that is demanded of us, the psychological maturation, this development of consciousness is a form of evolution. For evolution is of both bodies and minds, of matter and consciousness.[21]

As one ascends the ladder of power and influence, one's actions have greater impact. By splitting the atom and developing weapons of mass destruction, our species has accrued more power and influence. This affords us the opportunity to see as never before that what we do has impact. We can see as never before that we are the architects of our future. By recognizing our status, perhaps we will realize the need to make conscious choices. Teilhard de Chardin said, "Evolution is an ascent toward consciousness."[22] Likewise the development of consciousness is a form of evolution. If we as a species can genuinely absorb this reality, we will be empowered as never before. We will be reaching toward our highest human potential.

CONCLUSION

In this chapter we have written at length about anxiety. Perhaps the greatest anxiety about the predicament in which we find ourselves is the uncertainty about our future. We don't know how all this will turn out. The development of weapons of mass destruction has placed the human race at a crossroads. One road could lead to civilization's self-destruction, and the other road could lead to civilization's maturation.

These diametrically opposed paths were articulated by two prominent scientists at the time of the first atomic bomb explosion at Alamogordo, New Mexico, on July 16, 1945. J. Robert Oppenheimer, father of the atomic bomb, expressed the pessimistic view when he quoted Vishnu from the *Bhagavad-Gita* and said, "I am become death, the destroyer of worlds."

Albert Einstein, expressing the more optimistic view, is reported to have responded by saying that now all men are brothers.

These two voices reflect a dichotomy that exists in the internal environment of each of us. We are faced with a choice as to which voice will prevail. Will our species move down the path toward annihilation and ecological destruction, or proceed to a higher level of evolution?

NOTES

1. Elizabeth Kubler-Ross, *On Death and Dying,* New York: Macmillan, 1970.
2. Robert Jay Lifton, "Beyond Psychic Numbing: A Call to Awareness," *American Journal of Orthopsychiatry,* Oct. 1982, 52(4).
3. Goldenring and Doctor, "Teen-age Worry About Nuclear War: North American and European Questionnaire Studies," *International Journal of Mental Health,* (1986) 15, p. 72.
4. American Psychiatric Association Task Force Report 20, "Psychosocial Aspects of Nuclear Developments," Washington, D.C.: 1982.
5. Greg Diamond and Jerald Bachman, "High School Seniors and the Nuclear Threat 1975-1984: Political and Mental Health Implications of Concern and Despair," *International Journal of Mental Health,* (1986) 15, p. 210.
6. T. Solantaus, M. Rimpela, and V. Taipale, "The Threat of War in the Minds of 12-18 Year Olds in Finland," *Lancet* (1984) 8380(1) p. 784.

7. E. Sommers, S. Goldberg, D. Levinson, C. Ross, and S. LaCarbe, "Children's Mental Health and the Threat of Nuclear War: A Canadian Pilot Study" pp. 61-93, in *Impact of the Threat of Nuclear War on Children and Adolescents,* T. Solantaus, E. Chivian, M. Vartanyan, and S. Chivian, eds. Boston: International Physicians for the Prevention of Nuclear War, 1981.

8. P.O. Holmborg, and A. Bergstrom, "How Swedish Teenagers Think and Feel Concerning the Nuclear Threat," pp. 170–80 in *Impact of the Threat of Nuclear War on Children and Adolescents,* T. Solantaus, E. Chivian, M. Vartanyan, and S. Chivian, eds. Boston: International Physicians for the Prevention of Nuclear War, 1981.

9. Eric Chivian, et al, "Soviet Children and the Threat of Nuclear War," *American Journal of Orthopsychiatry,* Oct. 1985, 55(4).

10. Eric Chivian, John P. Robinson, Jonathan R.H. Tudge, Nikolai P. Popov, and Vladimir G. Andreyenkov, "American and Soviet Teenagers' Concerns about Nuclear War and the Future," *New England Journal of Medicine,* Aug. 18, 1988, 319(7).

11. George Kennan, *Boston Globe,* Nov. 29, 1981, p. A15.

12. Ronald Reagan, "Image of Enemy," *San Francisco Chronicle,* June 1, 1988, p. 1.

13. Ronald Reagan, "Reagan Has Warm Words for US-Soviet Relations," *San Francisco Chronicle,* Dec. 9, 1988, p. 1.

14. Jerome Frank, "Prenuclear-age Leaders and the Nuclear Arms Race," *American Journal of Orthopsychiatry,*" Oct. 1982, 52(4).

15. Judith Vigna, *Nobody Wants a Nuclear War,* Niles, Illinois: Albert Whitman & Co., 1986.

16. Joseph Adelson, and Chester E. Finn, Jr., "Terrorizing Children," *Commentary,* Apr. 1985, pp. 29–36.

17. Robert Coles, "Children and the Bomb," *New York Times Magazine,* Dec. 8, 1985.

18. Barbara Tizard, "Problematic Aspects of Nuclear Education," *Harvard Educational Review,* Aug. 1984, 54(3) pp. 271–81.

19. William R. Beardslee, "Children's and Adolescents' Perceptions of the Threat of Nuclear War: Implications of Recent Studies." *The Medical Implications of Nuclear War.* The National Academy of Sciences, Washington D.C.: National Academy Press, 1986, p. 416.

20. Marian Wright Edelman, Speech to Physicians for Social Responsibility, National Meeting, Washington, D.C.: March 1988.
21. Roger Walsh, *Staying Alive*, Boulder Co: New Science Library, 1984, p. 81.
22. Teilhard de Chardin, in Roger Walsh (1984), op. cit., p. 81.

11

The Economic Cost of Ineffective Weapons

LLOYD J. DUMAS

INTRODUCTION

THE TITLE OF THIS CHAPTER RAISES TWO IMPORTANT QUESTIONS: WHAT ARE "economic costs" and in what sense are weapons "ineffective"? Every weapon in the vast arsenals of the world's nations can of course be used to effectively kill or destroy. Yet death and destruction are not ends in themselves, but means to the end of either carrying out or thwarting organized aggression. Weapons are, after all, tools with which to fight wars, and wars are fought for the purpose of dominating others or preventing others from dominating.

The greatest part of the destructive capability contained in present day arsenals is in the form of nuclear weapons, weapons whose enormous destructive power makes them more instruments of annihilation than tools of battle. They are ineffective—even unusable—in the sense that anything more extensive than highly limited use threatens the continued existence of aggressor and defender alike. And once they are put to use, once the clearcut dividing line between conventional and nuclear conflict has been crossed, there are no obvious, mutually agreed upon stopping points to prevent further escalation. Any prudent military planner would have to assume that escalation to general nuclear war is more likely than ceasefire under the chaotic and frightening conditions that would prevail once two nations with immense nuclear arsenals had crossed the nuclear threshold in the throes of war.

There is, however, no meaningful difference in kind or concept between the economic costs of spending on nuclear or conventional weapons. Are conventional weapons also ineffective? There is no doubt that they are more usable than nuclear weapons. Yet they have become much less effective than in the past, much less effective than most of us have yet realized. Since the end of World War II, there have been very few cases in which the clashing of two or more organized armies has decisively and finally settled a conflict. The Middle East, for example, has seen a series of wars whose result to date has been to keep the pot boiling, not to settle the many complex disputes that afflict that troubled region.

Furthermore, any conflict that draws nuclear armed nations into conventional war with each other contains the potential for escalation into general nuclear war. It is virtually unimaginable that any such nation, on the verge of being decisively defeated in a conventional war that threatened its own territory or sovereignty would surrender rather than use its nuclear weapons. And even limited use of nuclear weapons to break the back of a conventional thrust by its nuclear armed adversary is, as discussed above, likely to degenerate into general nuclear war—war that will accomplish nothing less than mutual suicide at best, if not the extinction of human society. Conventional weapons may still be relatively effective for internal repression, but their effectiveness in settling international disputes has become highly questionable, to say the least.

In a great apparent paradox of history, the ability of military power to accomplish many of its traditional objectives has become much more limited as its destructive capability has dramatically grown. Yet as they have become more and more sophisticated, the economic costs of these weapons have continued to increase even as their effectiveness has continued to decline.

Economists measure cost by sacrifice. The true cost of anything is reckoned to be the value of what must be forgone in order to obtain it. Money costs are a form of shorthand, a convenient way of representing the real tradeoffs that lie beneath. For example, knowing that a desk costs $500 immediately tells you that purchasing that desk requires you to give up the use of that $500 to purchase two easy chairs costing $250 each, 25 compact disks costing $20 each, or any other combination of goods and services adding up to a cost of $500. The real cost of the desk is not $500, but rather two easy chairs or 25 compact disks, or the real goods and services in any other comparable combination. True cost may also be reckoned in terms of the sacrifice of resources. For example, for an individual whose wage is $12.50 per hour, the desk could be said to cost 40 hours of his or her working life—time that could have been devoted to any of a wide range of alternative activities.

Stating cost in money terms may be convenient, but it is sometimes misleading. It is very important to bear in mind that a proper reckoning of cost requires looking beyond the veil of money values to the real sacrifice of resources and alternative goods and services. Furthermore, it is important to distinguish between short- and long-term costs. It is not enough to consider the more obvious immediate sacrifices required by a given course of action. This may greatly underestimate true cost by ignoring long-term processes set in motion by this course of action—processes that may impose much additional sacrifice over the long run.

This is certainly the case in evaluating the economic costs of military expenditures. To the extent that the real tradeoffs involved in military spending have been analyzed at all, they have ordinarily been viewed in terms of "guns vs. butter," i.e., in terms of the short-term sacrifice of consumer goods. Even those who have extended the analysis beyond this direct consumption tradeoff have still tended to stay within the realm of the short term. For example, a number of analyses have pointed out that military spending creates fewer jobs per dollar spent than virtually any other form of government expenditure—or for that matter than the private spending that would be generated per dollar cut from taxes and left in private hands.[1] Thus jobs forgone are part of the sacrifice imposed by military spending. Others have compared expenditures on military programs with various forms of "social spending" forgone, from food stamps to public housing.[2] These are meaningful comparisons. They do in fact capture an important element of the real cost of military expenditures. But it is the often neglected long-term economic cost generated when high levels of military spending are continued over long periods of time that is the central focus of this analysis.

Since the end of World War II, and especially since the 1950s, the United States has maintained a large military establishment supported by a large network of military-oriented industries, research laboratories, and "think tanks" that have come to be known as the military-industrial complex. Has this emphasis on military activity damaged the American economy? Or has all this military spending provided a stimulus to general economic growth, a spur to commercial innovation, a competitive advantage in access to needed resources and markets around the world?

THE STATE OF THE AMERICAN ECONOMY

In 1982, the United States experienced the worst recession since the Great Depression of the 1930s. Since that time, we have been told that a powerful five-year recovery has occurred, that the economy has been revitalized, that America is "back on the track." But the economy has not recovered. There has been no economic revitalization. What we have done is to pump up the surface of the economy by engaging in nothing less than a national orgy of debt. We have not returned to the sound base of efficient production that drove the spectacular growth of the American economy throughout much of its history, the kind of solid foundation that can support real economic growth and make it sustainable. The base of the economy continues to erode, and on top of that weakening base we have created a "bubble." The erosion of America's economic base portends continuing long-term decline; the "bubble" of debt raises the possibility of sudden collapse. The signs that the American economy is in trouble are clear and unmistakable.

Since the mid to late 1970s, the real income (purchasing power) and hence the standard of living of Americans has been falling. The real income of the average (median) family was 5.2 percent higher in 1973 than in 1985; the real income of the "median couple" (i.e., the combined income of one male and one female, each earning at the median income for their gender) was 5.4 percent higher in 1973 than in 1985.[3] Yet another measure, purchasing power earned per hour worked, was more than 2 percent higher in 1978 than in 1985.[4]

There have been hundreds of bank failures since 1981. A post-Depression record of 120 bank failures was set in 1985, a record that lasted only one year; there were 138 bank failures in 1986. In 1987, 184 banks failed, more than 50 percent greater than the record set only two years earlier.[5] At the same time, the interest-bearing public debt of the federal government was undergoing an explosive increase. At the end of 1980, this national debt stood at $906.4 billion, an enormous amount of money. But in the next five years, it more than doubled to $1,821.0 billion, and by year end 1986 was well over $2 trillion.[6] Cutbacks in "social programs" were overwhelmed by an extraordinary expansion of military expenditures combined with politically popular but fiscally unwise tax cuts. Thus, in its first five years, an administration that took office promising to balance the federal budget by 1984 instead ran budget deficits that **averaged** more than three times those of its immediate predecessor.[7] And so that administration added as much to the national debt in just five years as was added by *all* previous administrations over more than two centuries.

The explosion of national debt, combined with historically high real interest rates, has itself made the federal budget much harder to balance. In 1985, net interest payments on the national debt alone absorbed 38.4 percent of the revenue received from the federal income taxes paid by all individuals and families in the United States. In 1986, that figure rose to nearly 40 percent. In fact, net interest payments on the federal debt combined with government purchases of military goods and services—those two budget items by themselves—absorbed all of the federal income tax revenues collected from all American taxpayers, individuals, families, and corporations in 1986.[8]

But it wasn't just the national debt that was rapidly increasing. By the summer of 1985, the United States had crossed a historic boundary: in the midst of our "strong recovery," we became a net debtor nation internationally, for the first time in 70 years. At the end of 1980, the net debt of the rest of the world to the United States was over $100 billion but within five years the situation had been completely reversed—by the end of 1985 the United States owed the rest of the world a net debt of more than $100 billion. Two years later (year end 1987), the U.S. net international debt had soared to well over $400 billion.[9]

It is difficult to interpret declining real income, a rising tide of bank failures, an explosive increase in the national debt, and a deepening dive into international net debtor status as signs of a healthy and strong economy. But as serious as these problems are, it may be that the clearest and strongest sign of serious economic trouble is the large and expanding trade deficit. Every year from 1894 through 1970, the United States exported more than it imported. Beginning in the 1970s, these trade surpluses turned into deficits. In 1983, the merchandise trade deficit hit an all-time record of $67.1 billion. The next year saw the trade deficit leap to $112.5 billion, a spectacular one-year increase of nearly 70 percent.[10] By 1987, the deficit had grown by another 40 percent.

America's trade deficits reflect nothing less fundamental than an inability to produce products people want at a price they can afford to pay—a failure of American competitiveness. If goods "made in U.S.A." were produced efficiently, they would be attractive in price and quality to customers here and abroad, as they once were. Exports would be higher and imports lower, and trade surpluses rather than deficits would still be the order of the day.

An economy that cannot produce efficiently cannot hope to continue to provide the high and rising material standard of living to which Americans have become accustomed. The decline of productive competence is about as basic an economic problem as there can be. That this is the key source of our troubles is a bitter pill to swallow for a nation that always prided itself on the spectacular efficiency of its industry. We don't want to believe it. So we keep trying to get ourselves off the hook by offering alternative explanations for our plight.

It has, for example, become common to accuse the Japanese of "dumping," i.e., selling products below cost. That sort of unfair competition, we keep telling ourselves, is the reason we are unable to compete. In the first place, the Japanese are far from the only nation outcompeting us in various world markets, though they are clearly a particularly striking case. Secondly, no producer can continue to sell below cost for an extended period of time in market after market. Dumping requires subsidy, and no one can continue to subsidize any producer indefinitely without having a source of wealth from which to pay the subsidy. In other words, even if they have engaged in some dumping from time to time, Japanese producers must be selling above cost somewhere to earn the money they must have to subsidize that dumping. And to argue that the Japanese government is subsidizing a policy of widespread dumping makes little sense. Where is the source of wealth from which the government is able to finance these subsidies? And, in any case, what is the point of continually selling your products below cost everywhere? They may well be selling below *our* cost in market after market, but that is not dumping; it is effective competition.

A strong dollar has also been blamed for our inability to compete. When the dollar is expensive relative to foreign currencies, U.S.–made products become very costly to foreigners while foreign products become relatively cheap to Americans. Indeed, a strong dollar does worsen our competitive position, but the dollar has been quite strong only since the early 1980s. The trade deficit was a serious problem for about a decade before the dollar began this rise. Furthermore, the dollar was so weak in the early 1970s that it was officially devalued in 1971, the year the trade balance began to turn against the United States. And the dollar has dropped sharply in the last few years as trade deficits continued to rise.

OPEC's high oil prices have been another popular explanation for the trade deficit, since the United States is a major importer of oil. High oil prices have aggravated the problem. However, the OPEC price rises began in late 1973, whereas the U.S. trade deficits began two years earlier, in 1971. In the past few years, oil prices have been stable or declining, while the trade deficits have continued to climb. Finally, if the United States had imported and exported only manufactured goods in 1984, it still would have easily broken the all-time trade deficit record set in 1983. In fact, counting *all* of the U.S. exports and *none* of its imports of oil (or other raw materials) that year, 1984 would still have seen a multibillion-dollar trade deficit.[11]

It has also been argued that high U.S. wages are the reason that domestic producers are unable to compete; that high rates of pay in the United States raise labor costs, creating a serious competitive disadvantage for American industry. While it is true that wages are high in the United States relative to those paid in much of the rest of the world, it has been less and less true over time. Wages in many of the nations with which we compete have been rising faster than those paid here. In 1980, "hourly compensation costs" to employers for production workers in manufacturing in the United States were not the highest in the world; they were not even close. Eight nations ranked above the United States, four of them by more than 20 percent.[12]

Even the "cheap Japanese labor" we have been so fond of blaming for Japan's competitive advantage is not so cheap anymore. Stagnating and even falling American wages combined with the recent drop in the international value of the dollar have brought the average industrial wage in Japan to a rough equivalence with that in the United States. For much of the period of continual trade surpluses, the difference in the cost of labor between the United States and the rest of the world was greater than it has been in the period of our continual trade deficits. This would not have been the case if high American wages were the primary reason for our competitive woes.

No, there is more to this competitiveness problem than Japanese dumping, a strong dollar, crude-oil prices, or high American wages. The problem is deeper, more structural: American industry has become inefficient. And so it has been steadily losing its ability to compete. But why has U.S. industry become so inefficient? What has turned the American industrial powerhouse into a faltering economic machine? And why has it been so difficult to see this deterioration or recognize just how fundamental its sources have been?

MONEY VALUE AND ECONOMIC VALUE

There are many goals that individuals and societies try to achieve, many needs that they seek to meet. Economic well-being is only one of these, albeit one that is often considered particularly important. Attainment of these goals, satisfaction of these needs, often requires or is greatly facilitated by the availability of particular goods and services that typically have money values in the form of prices or costs associated with them. For example, many of us turn to organized religion to satisfy a need for spiritual or moral guidance. The availability of church buildings and prayer books and the services of ministers facilitate organized religion's ability to help us in this quest. There are construction and operating costs associated with the buildings, prices that must be paid for prayerbooks, and salaries paid to ministers.

The economy is the part of the society that provides the material standard of living, that enables us to meet material needs. It therefore makes sense to define the economic value of a good or service by the extent to which it contributes to material well-being. It is clear from the example given above that not all goods or services with a money value contribute directly to material well-being. Some, such as the churches, serve other social goals. Thus, money value and economic value are not the same thing.

There are basically two types of goods and services that contribute to the material standard of living. Consumer goods (such as TV sets, automobiles, and haircuts) contribute to the material well-being of the population directly and immediately. In fact, when we say that someone has a high standard of living, what we ordinarily mean is that he or she has access to a substantial quantity and quality of consumer goods. Producer goods (such as metalworking machines, blast furnaces, and architectural services) also contribute to material well-being, though indirectly, by enhancing the economy's capacity to produce. Since both of these categories of products add to material well-being, they have economic value. Because they contribute to the central purpose of the economy, the activity of producing them can be referred to as "contributive." The activity of producing those goods and services that do not contribute to this economic purpose can correspondingly be called "noncontributive."[13]

It is important to emphasize that "noncontributive" is intended to be a descriptive term, not a pejorative one. It means simply that the activity in question does not add to material well-being, that it is without *economic* value. There are many other forms of value associated with noneconomic human needs or social goals: spiritual, moral, political, status, and others. Saying that a given activity is not contributive economically is very different from saying that it has no value to society. To be sure, some activities might be noncontributive economically because they are useless, without any form of value whatsoever. But most will serve some personal or social purpose and thus have some noneconomic value. These other forms of value are not somehow inferior to economic value; merely they are different.

Because economists as well as the general public tend to treat money value as synonymous with economic value, it is easy enough to overlook the fact that contributive and noncontributive money-valued activities have very different impacts on the functioning of the economy. On the empirical side, a measure like Gross National Product (GNP)—the total money value of goods and services produced in a nation in a year—is taken as a measure of total economic value generated, of the nation's level of material well-being. But because it confounds money value and economic value, a high or rising GNP alone tells us little about the condition of the standard of living, and hence the presence or absence of real economic prosperity. A billion dollars spent on housing and a billion dollars spent on public monuments will add exactly the same amount to the GNP.

On the theoretical side, both of the dominant schools of macroeconomic thought—which concerns itself with the functioning of the economy as a whole—focus attention on the size of an aggregate of money value: in Keynesian theory, it is the dollar volume of total spending in the economy; in monetarist theory, it is the money supply. From this point of view, a dollar is a dollar. For example, Keynesians would argue that if the government wants to stimulate the economy, it can do so by spending more money. It makes no macroeconomic difference whether the money is spent on roads, education, health care, or weapons. Yet, particularly over the long term, the impact of these different forms of spending on the functioning of the economy will be very different. Spending on contributive activities will help the economy to flourish, and spending on noncontributive activities will be an economic burden.

Although not all activities that produce goods and services have economic value, they all do have economic cost. The labor, machinery, fuels, and other productive resources they use could be used to produce goods and services that do have economic value. As discussed earlier, the real cost of any activity is reckoned in terms of sacrifice. Thus, the real economic cost of producing noncontributive goods can be measured by the economic value of those contributive goods forgone, goods that could have been produced with the resources that were instead used for noncontributive production.

The confusion of money value and economic value is one of the key reasons why it has been more difficult than it should be to see the deterioration that lies beneath the debt-inflated surface of the U.S. economy. As long as the GNP was growing—as it did at quite a rapid pace in 1983, for example, while the nominal rate of unemployment was dropping—we could believe a strong recovery was under way, even as the base of the economy continued to decay.

The fact is that an expansion of money value can have dramatic short-run effects whether or not it represents a real expansion of economic value. And economic theories that stress the short run, habits of business that focus attention on this quarter's (or at most this year's) profits, and the tendency of politicians to look to the coming election combine to predispose us to behave as though we live in a short-run world. This myopic worldview makes it hard to appreciate the enormous difference in the economic impacts of contributive and noncontributive activities, since that difference can appear minimal in the short run. Over the long run, however, the difference will make itself painfully obvious.

REAL ECONOMIC GROWTH, COMPETITIVENESS, AND THE STANDARD OF LIVING

The vast majority of the U.S. population earns the largest part of its income in the form of payment for labor, i.e., wages and salaries. Relatively few derive their income mainly from profit, interest, or rent. Therefore, in order to have widespread improvement in the standard of living, it is necessary for wages and salaries to rise over time. The problem is that rising wages push up labor costs of producers (typically the largest component of their costs), creating pressures for them to raise their prices. But as this cost-based pressure causes prices to rise, the rising prices wear away the purchasing power of the higher wages. They also tend to make products less attractive to customers than the products of any competitors that have not been pressed to raise prices. Thus, the key to maintaining the competitiveness of producers *and* achieving real improvements in the general standard of living is not to hold wages down, but to find ways of offsetting the cost-push pressure created by rising wages. That means increasing the efficiency of production—getting more output from the same workforce.

Though many things affect the efficiency of production (such as the quality of management and motivation of workers), two of the most critical are the advance of production technology and the availability of a sufficient quantity and quality of machinery, equipment, and other production capital. Given more and better tools with which to work and improved techniques of production, it is possible to increase the quantity and quality of output per worker. High and rising wages and salaries can therefore be paid without pushing labor costs *per unit of output* proportionately higher. For that matter, rising costs of fuels and raw materials can also be offset by increases in productive efficiency. Product prices can thus be held relatively low and product quality high even as wages and salaries rise. The end result: products that are highly competitive in price and quality, along with a broad-based improvement in the general standard of living.

However, if insufficient capital is available to producers and the stream of innovation is slowed, the rise in output per worker will be held back. Higher wages and salaries (and other input costs) will not be as effectively offset by increases in output per worker; prices will rise and product quality will fall relative to foreign competitors. The result: lost export markets abroad as foreign consumers turn to lower-priced, quality products produced elsewhere; lost markets at home as domestic consumers turn to imported goods for the same reasons; growing trade deficits, chronic unemployment and underemployment; declining real income; and hence falling living standards, as either inflation outruns wage increases or competitive pressures force wage concessions.

THE ARMS RACE AND THE STRUCTURAL DECAY OF THE AMERICAN ECONOMY

For the past 40 years, in the service of a large and continuing arms race, the United States has diverted from the civilian economy a substantial fraction of both critical capital resources and the resources that drive the progress of technology—the nation's engineers and scientists. It should therefore come as no surprise that the nation has in the process reversed its accustomed pattern of continuing improvement in the standard of living, or that it has hobbled the ability of U.S. producers to compete in domestic and world markets.

The diversion of capital to the military sector is discussed at some length in the chapter by Seymour Melman. I will only add that the total reported value of all physical capital directly owned by the military in 1983 was nearly 50 percent as large as the book value of the total capital stock owned by all U.S. manufacturing establishments combined.[14] It is hard to imagine how so substantial a preemption of the nation's capital could occur without retarding the improvement of productive efficiency.

At least 30 percent of America's pool of engineering and scientific talent has also been diverted to military-oriented activity over a period of three to four decades.[15] Because of its high salaries and other benefits (such as a free-spending environment that allows extensive access to state-of-the-art facilities and equipment), the military sector is able to attract not only large numbers of engineers and scientists, but also a disproportionate share of the graduates of the nation's premier science and engineering educational institutions. Thus, this internal "brain drain" has both quantitative and qualitative dimensions.

With so much of America's technological talent engaged in military-oriented research and development, the rate of progress of military technology has accelerated rapidly over the past few decades. Yet at the same time, *and for the same reason*, the pace of civilian-oriented innovation has slowed dramatically. Thus, the ability of American producers to offset high and rising wages with improved techniques of manufacture began to evaporate.

From the mid- to late 1960s through the 1970s, the continued rise of wages along with the rising costs of energy and other needed inputs created a base for considerable cost-push inflation. As the prices of goods "made in U.S.A." rose higher and higher, domestic producers lost markets at home and abroad to foreign competitors, forcing cutbacks in U.S.–based production. Unemployment became chronic. That some of the disappearing high-wage jobs in manufacturing have been replaced by a surge of low-wage, often dead-end service sector jobs is hardly cause for celebration. While wage concessions and a fortuitous fall in world energy prices caused inflation to abate briefly in the 1980s, trade deficits continued to rise and the standard of living continued to decline.

The U.S. economy has been undermined at the key historic source of its strength—the power of American innovation—not by some external enemy, but as an unintended result of our unquestioned belief that an ever larger stockpile of increasingly sophisticated weapons is the definition of security. We have done it to ourselves.

SPINOFF

It has often been argued that emphasis on military R&D stimulates civilian technology, rather than interfering with its progress; that technological progress has always been most rapid in periods of war or preparation for war. Technological advances made in the service of military objectives "spin off" to civilian application, it is argued, spurring the process of civilian invention and innovation. High levels of military R&D thus convey an advantage to commercial producers, not a penalty.

There is little question that such spinoff exists, or even that it has sometimes been of considerable value. But that is not the real question. From the point of view of the economy, the real question is which of two clearly contrary effects is stronger: the "spinoff effect" that tends to stimulate civilian technology, or the previously discussed "brain drain effect" that tends to retard it.

If spinoff were the more powerful effect, nations heavily engaged in military R&D would have a considerable advantage in developing civilian technology over nations that do comparatively little military R&D. If this were true, nations like the Soviet Union and the United Kingdom, which do a great deal of military research, would be in the vanguard of civilian technological progress, while nations like Japan and West Germany, which do relatively little military R&D, lag far behind. But of course, it is just the opposite.

Furthermore, if spinoff were the dominant effect, the huge amount of resources poured into military R&D in the United States over the past few decades would have made the eighties a golden age of American civilian technology. Instead it has been clear for more than ten years that the relative rate of civilian technological progress in the United States has been seriously reduced. The sorry state of American civilian technological progress has been widely chronicled in such popular journals as *Business Week*, with stories bearing titles like "The Breakdown of U.S. Innovation" and "Vanishing Innovation."[16] It has been documented in reports of the governing board of the National Science Foundation.[17] The 1985 report of the Commission on Industrial Competitiveness, specially appointed by President Reagan and heavily weighted with executives from military industry, also strongly emphasized the problems of American commercial innovation.[18] In short, it is clear that there is not nearly enough spinoff to compensate for the effects of the sustained diversion of the large numbers of engineers and scientists who have been drawn into military R&D.

Proponents of the spinoff argument typically cite as examples of spinoff such advances as radar or early computer technology, which originated during or soon after World War II. In that era, military and civilian technologies were much more similar than they are today. The technological gap between a forties bomber and its contemporaries in civilian aviation, for example, was minuscule as compared with the gap between a

B1 bomber and a Boeing 747. The striking divergence in recent years has sharply reduced the extent to which military R&D generates usable spinoff to civilian technology.

Furthermore, the capability for high performance (maximum possible speed, maneuverability, range, accuracy, etc.) under extreme conditions (of vibration, shock, heat, cold, humidity, dryness, etc.) demanded by present military forces, combined with minimal attention to cost, generates technology that has little relevance to viable, cost-sensitive civilian commercial application unless there is considerable additional development. In other words, it is not possible to capitalize on what limited spinoff potential does exist without the efforts of a considerable pool of engineers and scientists. With so many technologists drawn to military R&D, there may simply not be enough engineers and scientists working in civilian R&D to facilitate even the limited spinoff available. Increasingly, the spinoff from American military technology that does occur comes out of Japanese industry.

In addition, the secrecy that usually surrounds military R&D programs inhibits the free flow of information and collegial interaction so vital to technological advance. A civilian application cannot spin off from a military technology to which civilian researchers have no access.

REBUILDING A PROSPEROUS AND SECURE AMERICA

The military system does not function according to the principles of free-market capitalism. It is a command sector, asserting and enforcing its claim to the nation's productive resources through the special political priority it has been accorded. It continues to divert a substantial fraction of the society's technological talent and enormous amounts of physical and financial capital, undercutting the ability of American producers to operate efficiently and thus stay competitive with foreign producers. This diminution of competitiveness has brought with it an ongoing deterioration in material well-being. This generation of Americans may be the first to face a lower standard of living than its parents, but unless national priorities are rearranged, it will surely not be the last.

Removing a substantial part of the heavy military burden on the economy could open the door to an industrial renaissance unlike anything seen since the beginnings of the Industrial Revolution itself. Reducing the military drain would free huge quantities of technological talent and capital for productive use. But merely freeing these resources from noncontributive activity is not enough. Care must be taken to see that they are reconnected to contributive work in the civilian economy.

The extreme performance, cost-insensitive world of military R&D and production is light-years away from the good performance, highly cost-sensitive world of civilian R&D and production. People whose experience has been primarily in the former cannot function effectively in the latter without substantial retraining and reorientation. Facilities and equipment that have been optimized for military R&D and production have to be modified if they are to be effective in contributive civilian activity. This cannot be simply left to the market, which will not bridge the huge gap between the training of military sector personnel and the needs of the world of civilian commercial business.

The process of transferring people and facilities from the military sector to the civilian sector is called "economic conversion." It can be carried out smoothly and efficiently only if it is carefully planned in advance, in a highly decentralized fashion, facility by facility (see chapter 12).

Advanced, decentralized planning for conversion is a down-to-earth, practical policy. It assures those whose livelihood currently depends on the military sector that cutbacks in military-oriented activity do not threaten their personal economic security. That should help to remove one of the most politically important internal obstacles to reversing the arms race—the pressure put on the Congress by constituents who are economically dependent on the continued flow of military dollars.

But economic conversion is also a key to rebuilding the nation's economy. The real problem is not that too many critical resources have been engaged in the military-serving sector; it is that they have *not* been engaged in contributive civilian activity. It is not what they have been doing, but what they haven't been doing, that has caused the structural economic decay. If they are removed from the military sector to the unemployment lines, they will still not be engaged in the kind of contributive civilian activity that is vital to rebuilding the American economy. They must be reconnected to such activity, not merely removed from the service of the military.

We live in a troubled world, a world of nations that have not learned how to get along together very well. In such a world, a strong military may be a necessity, but an obsession with weapons is not. We cannot keep ourselves secure by obsessively building more and more weapons that are less and less effective. In so doing we can only tear our economy apart, even as our security declines.

We can provide for the military component of the nation's security without crippling its economy. But to do so, we must stop building every weapons system the Pentagon and its industrial suppliers can dream up. This requires a more critical analysis of which military programs are truly vital to national security, and which are driven by political patronage or military fantasy. Stemming the flood of dollars to the Pentagon will leave the military leaner and tougher at the same time that it frees critical resources for the sorts of activity needed to rebuild America's economic strength. In many ways, it is on the strength of our economy that our present security and our future prospects depend.

NOTES

1. See for example R. Bezdek, "The Economic Impact—Regional and Occupational—of Compensated Shifts in Defense Spending," *The Journal of Regional Science*, vol. 15, no. 2, 1975; Employment Research Associates, *Military Spending: An Occupational Hazard*, Lansing, Michigan: Employment Research Associates, 1984; and M. Anderson, M. Frisch, and M. Oden, *The Empty Pork Barrel: The Employment Cost of the Military Buildup, 1981-1985*, Lansing, Michigan: Employment Research Associates, 1986.
2. See for example S. Melman, *The Permanent War Economy*, New York: Simon & Schuster, 1974, app. 5, pp. 358–69; and R.L. Sivard, *World Military and Social Expenditures, 1987-88*, Washington, D.C.: World Priorities, 1987, p. 35.
3. *Economic Report of the President* (January 1987), Washington, D.C.: Government Printing Office, 1987, app. B, table B-29, p. 278.
4. Ibid., table B-43, p. 294.
5. For the data through 1985, see L.J. Dumas, *The Overburdened Economy*, Berkeley: University of California Press, 1986, fn. 9, p. 273; the 1986 and 1987 data were provided by telephone by the Financial Structure Section, Division of Research and Statistics, Federal Reserve Board, Washington, D.C. In addition to 184 actual bank failures in 1987, there were 19 cases of "open bank assistance," for a total of more than 200 FDIC interventions.
6. *Economic Report of the President* (January 1987), table B-81, p. 340.
7. Ibid., table B-73, p. 331.

8. Ibid., table B-74, p. 333.
9. J. Faux, "Getting Rid of the Trade Deficit: A Cheaper Dollar Is Not Enough," Briefing Paper, Washington, D.C.: Economic Policy Institute, March 1988.
10. *Economic Report of the President* (January 1987), table B-99, p. 358.
11. Ibid., table B-102, p. 362.
12. S. Melman, *Profits Without Production*, New York: Alfred Knopf, 1983, p. 309.
13. For a more complete development of these concepts, see L.J. Dumas, *The Overburdened Economy*, chap. 4, pp. 52–77.
14. Ibid., p. 221.
15. Ibid., p. 208–11.
16. *Business Week*, "The Breakdown of U.S. Innovation" (26 February 1976) and "Vanishing Innovation" (3 July 1976).
17. National Science Foundation, National Science Board, *Science Indicators, 1974* and *Science Indicators, 1978*, Washington, D.C.: Government Printing Office, 1975 and 1979 respectively.
18. President's Commission on Industrial Competitiveness, *Global Competition: The New Reality*, Washington, D.C.: Government Printing Office, 1985.

12

Why Economic Conversion?

Seymour Melman

ECONOMIC CONVERSION IS THE PLANNING AND IMPLEMENTATION OF A CHANGE-over of factories, research laboratories, bases, and allied facilities from military to civilian application. After enduring a warlike economy for more than 40 years the American people need economic conversion planning for the following vital purposes:

- To make possible the physical reconstruction of U.S. industry, infrastructure, and environment that have been depleted by the war economy.
- To create confidence in the ability to carry out the production cutbacks that are a core part of an agreed disarmament process.
- To make disarmament an opportunity, rather than a penalty, for those whose working lives will have to be converted to civilian skills and tasks.

Economic conversion is dramatically different from what was called *reconversion* at the close of World War II. At that time factories that had turned from civilian to military production for the duration of the war reverted to their former activities. They were able to reapply the old machines and often the former workforce to the production system that had been there before, turning out familiar products, getting raw materials from familiar suppliers, selling to familiar markets.

In recent decades the war-making institutions have mobilized the largest labor force under one management. They also have the largest research and development staffs subject to one management. The R&D

sections of the Department of Defense and its allied agencies consume from 70 to 75 percent of the total R&D budget of the federal government.

We are indebted to President Eisenhower for calling attention, in his farewell address of January 7, 1961, to the fact that during every year of the previous decade, the fresh capital resources made available to the Department of Defense exceeded the cumulative net profit of all American corporations. That relationship has continued to the present day. Hence, the war-making institutions have been wielding the largest finance capital fund in the American economy. From 1946 to 1980 the budgets of the Department of Defense totalled $2,001 billion. The budgets executed and projected for the period 1980–88 were $2,089 billion. If control of finance capital is a means of exerting decision power—economic, political, and other—then it is only too clear that the war-making institutions of the United States have enjoyed, and do enjoy, an unprecedented measure of decision power.

COSTS TO SOCIETY

A modern military budget is a capital fund. Capital in an ordinary industrial enterprise is of two sorts: fixed and working. Fixed capital is the money value of the land, buildings, machinery. Working capital is the money value of all other resources that must be brought to bear to set the enterprise in motion. A modern military budget, as funds are applied, provides precisely the set of resources that are conventionally identified as fixed and working capital. And they do that on behalf of a set of products that have a unique characteristic, which is that they contribute nothing to ordinary consumption (i.e., to the standard of living), nor can they be used for further production.

Because of that peculiarity of the military product, the aggregate cost to the whole community of the military enterprise is almost entirely overlooked in our economic textbooks. The cost to the whole society is *first* the money-measured direct resources allocated to the military function. A *second* cost is measured by the civilian use-value, goods and services for consumption or for production, that are forgone when the resources are used for military products. For the whole community, that value forgone is approximately equal to the cost of the direct imputs from the military budget. There is a *third* cost: when capital resources are used for new civilian production on a continuing basis, there is usually improvement in the design of the means of production. They become more competent, more efficient, more productive. Economists call this an increase in the marginal productivity of capital. But when capital resources are used for purposes that cannot be applied to any further production, that increment of productivity of capital is lost forever.

There is yet a *fourth* kind of social cost to the whole society. A large part of the knowledge that is generated on behalf of the military has unique application to military technology alone and is not relevant elsewhere. So the possibility of the knowledge being used elsewhere (which would be likely if the R&D were carried out for civilian purposes) is forgone. A *fifth* kind of cost, although unmeasurable, is quite important. As a large labor force is trained for the military, it becomes habituated to working not with an eye to minimizing cost but to maximizing cost, and to offsetting that maximized cost with maximized subsidies. Such practices are possible only when carried out by a ward of the state, with almost unlimited resources made available by tapping the income of the whole society. A labor force so trained develops what Thorstein Veblen genially called a trained incapacity—in this case for civilian work. To my knowledge no one has yet attempted an aggregate measure of the social cost of the military in the categories I have identified.

HOW THE MILITARY DEPLETES THE CIVILIAN SECTOR

Returning to the basic idea of the military budget as a capital fund, it is important to compare that budget with another capital fund, the Gross Domestic Fixed Capital Formation. That simply means the money value of new civilian capital items put in place in a given year—school buildings, factories, machinery, roads, waterworks, libraries, and the like. From data on national income assembled by the staff of the United Nations one can calculate the capital fund made available for the military in a number of countries for a given year as a ratio per hundred dollars (or francs or yen) of capital represented by new civilian capital formation (whether privately or publicly funded). For the last year of such available data, 1983, here are the ratios: in the United States, for every $100 of civilian gross fixed capital formation, $40 was separately expended for the military; in the United Kingdom the ratio was 31; in France 16; in Sweden 12; in West Germany

13; in Japan 3.3. For the U.S.S.R. there are no official data. My estimate, since no proper data are offered, is 66.

These relationships are informative. The very low ratio of military to civilian use of capital in Japan and West Germany is a fundamental clue to the modernity of industrial plants and equipment, to the rapidly developing civilian technologies in those countries, and to the ability of those countries to sustain a rapid rise in wage rates together with competitive competence. The tradeoff—which economists call "opportunity cost"—between the military and civilian uses of capital-type resources is highly visible in the decay of infrastructure in the United States. In 1981, public works planners for state governments published a report on "America in Ruins"—an unprecedented concept. On September 27, 1982, the editors of *U.S. News and World Report* estimated that an outlay of not less than $2,500 billion was needed to repair the damaged infrastructure. But by 1988, such resources were still not available for these purposes, and it is only prudent to assume that the deterioration has become more severe and that a 1988 estimate of the cost of repair exceeds $3,000 billion.

That sum should be contrasted with the military budgets of the United States, 1947–87. In dollars of 1982 value they equalled $7,620 billions (Dept. Of Defense, Comptroller's Office).*

A further estimate in this area is truly startling. It is based on the U.S. military budget as planned to 1988. For the military there is a five-year budget plan—five-year plans being as American as apple pie when prepared for the Pentagon. I compared that 1988 budget with an estimate, derived by simple statistical extrapolation, of the Gross Domestic Fixed Capital Formation in the same year. And the U.S. ratio I arrived at for 1988 is 87:100. If the present trend of military buildup continues, if the intensive use of capital for the Pentagon prevails, it is entirely likely that the competence of U.S. industry will be irreparably destroyed, that the quality of the civilian production system will be so degraded as to be an important factor in reaching what I describe in *Profits Without Production* (Univ. of Pennsylvania Press, 1987) as a point of no return.

Coupled with concern about the military budget as a capital fund is concern about its influence on productivity growth. From 1865–1965 the United States paid the highest industrial wages in the world, and its industries were able at the same time to turn out products of a quality and

* Some of the details of this immense tradeoff were reported in the *New York Times* of July 26, 1981, in an article titled "Looting the Means of Production."

at a price that made them salable both domestically and abroad. That was possible because high wages were offset by progressive mechanization of work and improvement in the organization of work. As a result, there was rising output per man-hour, an increase in productivity. But after 1965, following a period of expanding military budgets, productivity growth of the United States dropped sharply. According to a December 31, 1984, report of the U.S. Bureau of Labor Statistics on trends of manufacturing productivity and labor cost, 11 countries, including Canada, Japan, Belgium, Denmark, France, Germany, Italy, the Netherlands, Norway, Sweden, and the United Kingdom, during the period 1960 to 1983, all had higher average rates of productivity growth than the United States, whose average annual increase was only about 2.6 percent.

The ratio of military to civilian use of capital therefore has profound consequences, for mechanization of work and modernity of the industrial plants are the basic requirements for substantial improvement of productivity growth. But as cost-minimizing was replaced in the military industry by cost-maximizing, the classic process whereby endless detailed improvements were made in manufacturing, productivity was stalled. The mechanism of cost-minimizing in U.S. industry used to work as follows: the wages of labor, a major cost item, would rise for various reasons—market forces, union bargaining, whatever. The users of labor in cost-minimizing firms would move to offset the wage increase by mechanization of work, by better organization of work, by redesign of product. When this took place among the firms that produced industrial machinery, it meant that the wages they and their customers paid could rise more rapidly than the prices of their product-machinery. In turn, this meant that the users of machinery were being offered new, more efficient equipment at prices that were progressively more attractive. So a continuing process invited the broader, more intensive use of machinery in place of manual labor. The automatically derived effect was an improvement in the average productivity of labor.

In an earlier study, *Dynamic Factors in Industrial Productivity* (1956), I showed this pattern as characteristic of the United States, and of Great Britain until 1950. Now that pattern has been essentially transformed. Table 1 shows percent changes in average hourly earnings to industrial workers and percent changes in machine tool prices, 1971–78.

In Japan average hourly earnings to production workers rose 177 percent while machine tool prices rose only 51 percent. There is the classic pattern. The producers of machine tools contrived to offset the wage increase with internal plant efficiencies so that the prices of their product

	Worker Earnings/Hour	Machine Tools Prices
United States	+ 72	+85
West Germany	+ 72	+59
Japan	+177	+51

(Source: *Profits Without Production*, p.174)

Table 1

rose only 51 percent. The West German differential, 72 and 59, is smaller, but one that invites further mechanization of work. While U.S. average hourly earning rose as much as those in West Germany—72 percent—U.S. machine tool prices rose 85 percent, thus discouraging widespread mechanization of industrial work.

By 1978 the United States had the oldest stock of metalworking machinery of any industrialized country. This was due partly to the fact that the Department of Defense had become a major purchaser of machine tools and a major sponsor for machine tool R&D, thus causing cost-maximizing to infect the interior operations of the machine tool industry. The Pentagon's erratic bursts of production orders have been a major contribution to an unstable market for machine tools. There was also a general transformation in the emphasis of U.S. industrial management, away from competent production and toward making money by all manner of financial devices.

HOW A PERMANENT WAR ECONOMY COUNTERS PRODUCTIVITY GROWTH

There is a definable set of conditions that favors productivity growth. They are: cost-minimizing within the machinery-producing industries, cost-minimizing among machinery users, wages of labor rising faster than machinery prices, availability of finance capital at modest interest rates, R&D to innovate new means of production rather than financial devices, stable rates of production, a management oriented to production, and the availability of a competent surrounding infrastructure as a production support base. The effect of a military economy in its normal operation has run counter to every one of these requirements for productivity growth. As a result, American firms have become less able to offset cost increases from whatever source and therefore have become unable to supply even the domestic market at acceptable levels of price and quality.

By 1980, 27 percent of the automobilies in the United States were imports, as were 25 percent of the machine tools (by 1986 it was 49 percent), 15 percent of the steel mill products, 87 percent of the black-and-white TV sets, 47 percent of calculating machines, 22 percent of microwave ranges and ovens, 34 percent of integrated micro circuits, 24 percent of X-ray and other irradiation equipment, 74 percent of motion picture cameras, 51 percent of sewing machines (by 1984, 100 percent of the household type), 100 percent of the radios, 22 percent of the bicycles, 50 percent of shoes (by 1987, 86 percent).

Every one of these figures also means that the same percentage of domestic employment in those industries has disappeared. That is what accounts for the several million jobs lost in the U.S. industrial economy.

MISLEADING STATISTIC

There is a discrepancy between the national employment figures issued by the government and the details of life in particular industries, cities, and regions. An aggregation of figures obscures the diversity of experience in the U.S. economy and society. The military economy prospers; there are boom conditions in the industries and in the regions that seek to employ engineers and skilled workers for the Pentagon's needs. The evidence may be seen in job announcements in the newspapers of Los Angeles, Boston, and New York City. But there is no boom condition in Homestead, Pennsylvania, where the works of the U.S. Steel Corporation are blacked out and where the neighboring works of the Mesta Steel Company, once supplier of steel mill equipment to the U.S. steel industry, have been dismantled.

Again and again it is essential to differentiate among the published aggregate figures for economic activities. For example, on December 22, 1984, an AP dispatch reported that, according to the Commerce Department, orders for durable goods posted their biggest gain in more than four years—an 8.3 percent November increase. Next paragraph: more than half the gain was attributed to a 99.4 percent increase in the volatile military orders category. In other words, the headline reads "Durable Orders Surge 8.3 Percent," but the small type tells us more than half was in readily manipulated Pentagon orders.

THE HIGH-TECH INDUSTRIES

Depletion has occurred not only in the so-called smokestack industries, but also in the highest of all the high-tech worlds—the computer industry. An instructive example is one of the most popular, fastest selling, and important of computers, the IBM Personal Computer. On the back of the cathode ray tube box is prined "made in Taiwan" or "made in Korea." The label on the lower box says "made in the U.S.A." but the disk drive is made in Singapore or Hong Kong, and the chips, made in Japan, are also assembled abroad. The keyboard and power unit are made in Japan. What is left for production in the United States is the outer case and the assembly of the main working components. Indeed, for the R&D–intensive industries as a group, the 1980s saw a sharp reversal of position. In 1980 the high-tech industries group showed a balance of payments that was favorable for the United States. By 1986 that had become a net deficit.

This situation is troublesome not for some nationalist reason, or because of the desirability of economic autarchy, but because in order to live a community must produce. There is no theory or experience from which to infer that it is possible on any sustained basis for a community to have its production done for it by someone else.

PRODUCT UNRELIABILITY

There is a further problem for which the military economy is responsible. The military trains engineers and production workers to accept product unreliability. Military products have been made increasingly complex, to the point that the complexity necessarily reduces reliability. That happens because the reliability of a set of linked components is the product of the reliability of each of the separate components. Since no man-made device is ever 100 percent reliable, the coupling together of more and more components degrades the reliability of the mechanism or system.

It is normal in military experience that of 100 F-15s, the first-line fighter plane of the U.S. Air Force, 45 of them have, on the average, been in repair at any given moment. If a set of M-1 tanks are on maneuvers, they can be kept running only by heroic efforts of maintenance staffs, and by the immediate availability of spare parts and abundant fuel—3.86 gallons being required per mile of tank travel. On the antiaircraft Aegis cruiser, software is crucial to the operation of the computer-control systems for the weaponry involved, but the software has a high failure rate. And other parts of the system can be expected to function only 45 percent of the time. Such unreliability, which has become acceptable in the military, is completely unacceptable in civilian operation.

If a trolley car or subway goes out of service 45 percent of the time, if it fails frequently on the tracks, it will block the right of way; it will inconvenience a whole community; it will immediately come to the attention of many witnesses. There will surely be a great public outcry for relief from an intolerable situation.

When aerospace technology and modes of organization were applied to the production of trolley cars (as was done by the Boeing-Vertol Co. outside Philadelphia; the full details are in chapter 13 of *Profits Without Production*), it became clear why the military-industry approach won't work in civilian economy. Boeing-Vertol had to drop out of the trolley car business because it was confronted by massive lawsuits from the Massachusetts Bay Transit Authority. The trolley was handsome; it was a pleasure to ride in when it functioned, but it had a remarkable number of things that went wrong. The things that went wrong were built-in weaknesses, the result of the mode of organization and operation that is systematically pursued in military industry, and notably in the aerospace industry. That is one important reason why economic conversion, which includes retraining personnel from military to civilian competence, is essential.

One of the standard practices of the military industry is called concurrency. In the civilian economy, when a new product is projected, it customarily goes through a set of functional stages: the concept; research; something is designed; a prototype is used; the prototype is tested; faults are discovered; it is redesigned, retested—perhaps several times. Only after the prototype meets the desired performance standards is the order given to produce. Not in the land of the Department of Defense. There, concurrency is practiced—indeed, ordered—by the regulations of the

Secretary of Defense. Concurrency means carrying out all these steps at the same time. Then, when you have produced and used the product, you discover defects, whereupon you change the design for the items yet to be produced and retrofit those already made. This is probably the most expensive way of building anything ever formulated. As this becomes the habitual behavior of military engineers and managers, and to some degree production workers, they become incapacitated for functioning in civilian economy. This is why occupational retraining for economic conversion is not an outrageous idea. It is indispensable if the persons engaged in the military economy are to function competently in a civilian environment.

PLANNING FOR ECONOMIC CONVERSION

What are the strategic components of economic conversion? *First*, it must be planned well in advance. To establish an alternative civilian use for a military-industrial enterprise of size requires a lead time of at least two years. This period is required for selecting new products, for selecting new production equipment, identifying sources of raw materials, retraining the labor force. For firms (or divisions of firms) with an all-military background, the selection of new products is crucial. *Second*, economic conversion planning must be decentralized. If attempted from a national center, it will fail because it is not possible to oversee from a remote headquarters the myriad specific data that must be taken into account in a plan devised for a given enterprise. *Third*, the whole activity must be carried out without subsidy to the managements involved. A continuance of subsidy means a continuance of cushioning for what in the civilian realm would be gross incompetence. *Fourth*, it is essential that every military-serving factory and base be obliged by law to set up an alternative use committee—half of its members being selected by management, half by the working people (all grades)—with responsibility and authority for planning alternative products, production, and marketing systems for the plant, laboratory, or the base facility. This must be made mandatory because otherwise the managements involved, and even some of the engineers or production workers, would be massively reluctant to engage in such planning.

Fifth, managers and engineers must be retrained. Military-serving managers are expert in the politics and diplomacy of dealing with the Department of Defense; they couldn't sell a folding bed to Sears to save their lives. Engineers who have been trained to create products of escalating intricacy, without regard to cost, must be taught a new approach to product design that seeks cost-minimizing simplicity within the limits of reliable performance. Research and development must be disciplined so that people who have been accustomed to making lavish use of resources paid for out of the federal pocket will discover economical ways of doing their work in a civilian environment. Some of the production workers need to be retrained, but probably a smaller proportion than engineers and managers.

Sixth, economic conversion planning must also include funds for relocation, because parts of typically oversized administrative and engineering staffs will have to be moved to other locations. *Seventh*, even with the most thorough advance planning, there is bound to be a lag between the end of the military production and the start of civilian work. Some support to the working people will therefore be required.

What can ex-military-serving firms produce? For a start they can examine the list of manufactured products now being imported to the United States, asking: which of these products could be produced in our facilities; could that work be done at acceptable price and quality? There is also a challenge to compete against competent producers already in the field. Also, American industries will require vast capital investment; the same applies to American infrastructure. As noted earlier the crumbling infrastructure of the United States could be repaired at a price exceeding $2,500 billion. This defines the scale of potential markets. U.S. military budgets (spent and planned) 1980-1990 exceed $2,900 billion. A Labor-Management Infrastructure Task Force reported "A Consensus on Rebuilding America's Vital Public Facilities." (Oct. 1983) The study was coordinated by Arnold Cantor, assistant director, Department of Economic Research, AFL-CIO, and Russell S. Davis, vice president for strategic planning, Union Pacific Corporation; it stressed highways, bridges, and urban water supply and wastewater treatment facilities. Enormous new markets for expanding industry are defined here.

H.R. 101

At this writing the set of requirements for conversion planning that I have enumerated here are incorporated in the text of a bill proposed by Representative Ted Weiss, (D.,N.Y.) and supported by more than 60 colleagues. In 1989 it is designated H.R. 101, a bill to "facilitate economic adjustment of communities, industries, and workers through reduction or realignment of defense or aerospace contracts, military facilities, and arms export and for other purposes."

The importance of H.R. 101 is twofold. First, it outlines a coherent, urgently needed, constructive approach that is an economic alternative to the arms race. Second, politically, with planning for economic conversion both the government and the people will have fresh confidence for negotiating and implementing the reversal of the arms race.

With a combined address to disarmament and economic conversion we can strike at the heart of the single most dangerous political process of our time. That is the antagonistic cooperation whereby the war-making institutions of the United States and the U.S.S.R. have reinforced each other and generated larger budgets, larger military economies, larger stocks of more dangerous weapons, and in doing so have justified more centralism, more authoritarianism in each society. These processes lead, in both societies, to economic exhaustion and political/moral decay. I know of no single public issue that is as crucial for the survival of society as disarmament and economic conversion.

A SELECTED BIBLIOGRAPHY

H.R. 101, a bill introduced to the 100th Congress, 1st Session, January 28, 1987, by Representative Ted Weiss (D., N.Y.).

Irving Bluestone, "Problems of the Worker in Industrial Conversion," *Journal of Arms Control.* 1:3 (1963), 589–96.

California Public Policy Center, *Jobs from the Sun: Employment Development in the California Solar Energy Industry.* February 1978. 119 pp. From California Public Policy Center, 304 S. Broadway, Room 224, Los Angeles, CA 90013.

John R. Cambern, and David A. Newton, "Skill Transfers: Can Defense Workers Adapt to Civilian Occupations?" *Monthly Labor Review.* 92 (June 1969), 21-25.

Lloyd J. Dumas, "Economic Conversion, Productive Efficiency and Social Welfare." *Journal of Sociology and Social Welfare.* 4 (January–March 1977), 567–96.

Lloyd J. Dumas, ed, *The Political Economy of Arms Reduction: Reversing Economic Decay.* AAAS Selected Symposium 80. 1982. Westview Press, Inc., 5500 Central Ave., Boulder, CO 80301.

Lloyd J. Dumas, *The Overburdened Economy*. Berkeley & Los Angeles: University of California Press, 1986.

B. Curtis Eaton, "Do Defense Engineers Have Special Reemployment Problems?" *Monthly Labor Review*. 94 (July 1971), 52–54.

D. Elliott, *The Lucas Aerospace Workers' Campaign*. Young Fabian Pamphlet 46, November 1977. Civic Press Ltd., Civic St., Glasgow G4 9RH, Scotland.

D. Elliott, M. Kaldor, D. Smith, and R. Smith, *Alternative Work for Military Industries*. August 1977. Richardson Institute for Conflict and Peace Research, 158 N. Gower St., London NW1 2ND, England.

Suzanne Gordon, and Dave McFadden, eds., *Economic Conversion: Revitalizing America's Economy*. Cambridge, MA: Ballinger Publishing Co., 1984.

W.F. Gutteridge, "Problems of Conversion of Scientists and Technologists in the Event of Disarmament." 1967. Dept. of Languages and Social Science, Lanchester College of Technology, Coventry, England.

Index

Index

About the Authors

ANNE H. EHRLICH is a senior research associate in the Department of Biological Sciences at Stanford University. She has written or co-authored dozens of scientific and popular articles and eight books on environmental issues, including those associated with nuclear war such as nuclear winter. Since 1985 she has chaired the Sierra Club's Military Impacts on the Environment Committee.

JOHN W. BIRKS is Professor of Chemistry and Fellow of the Cooperative Institute for Research in Environmental Sciences (CIRES) at the University of Colorado, Boulder. As an atmospheric and environmental chemist, his past work has emphasized studies of stratospheric ozone depletion and the climatic effects that would accompany nuclear warfare. In 1982 he and Paul Crutzen published the seminal paper on the atmospheric effects of nuclear explosions that led to the "nuclear winter" theory. Dr. Birks is a founding member of the Sierra Club's Committee on the Military Impacts on the Environment.

THOMAS B. COCHRAN is a Senior Staff Scientist with the Natural Resources Defense Council (NRDC). He is director of the Nuclear Weapons Databook project, and he initiated and co-directs (with Jacob Scherr) the NRDC/Soviet Academy of Sciences Nuclear Test Ban Verification Project. He also is co-director of the Nuclear Program in NRDC's Washington, D.C. office. He has served as consultant to numerous government and nongovernment agencies on energy, nuclear nonproliferation and nuclear reactor matters.

LLOYD J. DUMAS is Professor of Political Economy and Economics at the University of Texas at Dallas, and formerly associate professor of Industrial Engineering at Columbia University. He is author of *The Overburdened Economy* (Univ. of California Press, 1989); co-editor and contributing author of *Making Peace Possible; the Promise of Economic Conversion* (Pergamon Press, 1989); and editor and co-author of *Reversing Economic Decay; The Political Economy of Arms Reduction* (Westview Press and AAAS, 1982). He has published more than 60 articles in books and journals in six disciplines, as well as in the popular press. From 1980 to 1983, he served on the Committee on Science, Arms Control, and National Security of the AAAS.

GERALD JACOB is Assistant Professor of Geography and Public Planning at Northern Arizona University in Flagstaff. He has worked as a policy analyst for the state of Utah's High-Level Nuclear Waste Office and for the Western Interstate Energy Board. His book, *Site Unseen: The Politics of Siting a Nuclear Waste Repository* was recently published by the University of Pittsburgh Press.

ANDREW M. KIRBY is Associate Professor in the Department of Geography and Regional Development at the University of Arizona, Tucson. He has written more than eighty articles and a dozen books on urban and political questions. His book *The Politics of Location* (1982) is a detailed examination of the conflicts that surround the siting of nuclear facilities and the transport of materials through communities.

GARY LAPID is Associate Clinical Professor at Stanford University, where he teaches Psychiatry and Behavioral Sciences. He also has a private practice of psychiatry in Palo Alto, California. Dr. Lapid is co-founder and chairman of Project Victory—a national campaign to bring Americans of diverse perspectives together for non-adversarial problem solving, vis-à-vis national security. He is co-author of *The Great Turning* (1989) and has written and lectured widely on psycho-social aspects of the nuclear arms race.

SEYMOUR MELMAN is Professor Emeritus of Industrial Engineering at Columbia University. His academic work has cut across economics and engineering to study the way social, mainly economic, factors determine the design and use of technology. For more than three decades he has written on the arms race and the problem of reversing it, economic consequences of military weaponry, and the feasibility of conversion from a military to civilian economy. His most recent books are *Profits Without Production* (1983, 1987) and *The Demilitarized Society* (1988).

BENNETT RAMBERG is Senior Research Associate at the UCLA Center for International and Strategic Affairs. Dr. Ramberg has been a research fellow at the Stanford Arms Control Program and Princeton University's Center of International Studies and has served as a foreign policy consultant to the U.S. Congress. He has authored and edited a number of books, including *Energy and Security in Industrializing States* (co-edited with Raju Thomas, 1989) and *Arms Control Without Negotiation: The Role of Independent/Unilateral Initiatives* (forthcoming).

DAN W. REICHER is a senior project attorney with the Nuclear Program of the Natural Resources Defense Council (NRDC) in Washington, D.C. Mr. Reicher has led a number of successful litigation and legislative efforts to compel the Department of Energy to comply with environmental laws at its nuclear weapons production facilities. He has also mounted successful challenges to inadequate agency rulemaking, including standards promulgated by the Environmental Protection Agency for the disposal of high-level radioactive waste. Prior to law school at Stanford, he served on the President's Commission on the Accident at Three Mile Island.

JASON SALZMAN is a Research Associate with the Nuclear Program of the Natural Resources Defense Council (NRDC) in Washington, D.C. Mr. Salzman has conducted research and lobbied in Congress on issues relating to nuclear weapons production. Prior to working at NRDC, he studied the antinuclear movement in New Zealand on an Arnold Fellowship.

S. JACOB SCHERR is a senior attorney with the Natural Resources Defense Council (NRDC) in Washington, D.C. Mr. Scherr is Co-Director of NRDC's Nuclear Program, including its project on the environmental impacts of U.S. nuclear weapons production. He has worked extensively on international environmental issues, including nuclear proliferation, hazardous exports, international development assistance, verifiability of a nuclear test ban treaty, and the environmental consequences of nuclear war. He led a successful NRDC lawsuit to force the DOE to prepare an environmental impact statement on its planned restart of a production reactor at the Savannah River Plant in South Carolina.

JAY B. SORENSON is a Professor of Political Science at the University of New Mexico and a consultant to Los Alamos National Laboratory. He has authored studies of the Russian Revolution, Japan and nuclear weapons, and articles on issues ranging from national security to the environment. A twenty-year member of the Sierra Club, he is prominent in his home state of New Mexico for his efforts to save mountains and wild rivers. He is currently a member of the Sierra Club's Military Impacts on the Environment Committee.

F.J. (RICK) VAN RHEENEN is in the private practice of psychiatry in Palo Alto and serves on the teaching faculty of the Departments of Medicine and Psychiatry at Stanford University Medical School. He has spent a good part of his adult life in the quest for cross-cultural understanding. After medical school at the University of Oregon he spent two years in the Peace Corps in Nigeria. While training at Stanford where he earned a Master's degree in Anthropology, he ran cross-cultural support groups for foreign students at the International Center there. Dr. Van Rheenen has served for the past three years as Co-President (with Gary Lapid) on the Stanford Chapter of Physicians for Social Responsibility.